5th International Scientific Workshop on Advanced Technologies of Materials Field-Assisted Consolidation 2016

AA006754

IOP Conference Series: Materials Science and Engineering
Volume 218

Moscow, Russia
28 - 31 August 2016

Editors:

Eugene A. Olevsky
Evgeny G. Grigoryev

ISBN: 978-1-5108-4419-3
ISSN: 1757-8981

Printed from e-media with permission by:

Curran Associates, Inc.
57 Morehouse Lane
Red Hook, NY 12571

Some format issues inherent in the e-media version may also appear in this print version.

Printed by Curran Associates, Inc. (2017)

For permission requests, please contact the Institute of Physics
at the address below.

Institute of Physics
Dirac House, Temple Back
Bristol BS1 6BE UK

Phone: 44 1 17 929 7481
Fax: 44 1 17 920 0979

techtracking@iop.org

Additional copies of this publication are available from:

Curran Associates, Inc.
57 Morehouse Lane
Red Hook, NY 12571 USA
Phone: 845-758-0400
Fax: 845-758-2633
Email: curran@proceedings.com
Web: www.proceedings.com

Table of contents

Volume 218

5th International Scientific Workshop on "Advanced Technologies of Materials Field-Assisted Consolidation"

28–31 August 2016, Moscow, Russian Federation

Accepted papers received: 27 June 2017
Published online: 5 July 2017

Preface

Papers

IOP Conf. Series: Materials Science and Engineering **218** (2017) 011001 doi:10.1088/1757-899X/218/1/011001

5th International Scientific Workshop «Advanced Technologies of Electromagnetic Field Assisted Consolidation of Materials »

Moscow, Russia, 29-31 August 2016

Organizers:

National Research Nuclear University «MEPhI»

Key Laboratory for Electromagnetic Field Assisted Materials Processing

Preface

These proceedings are a collection of the selected articles presented at the 5th International Scientific Workshop «Advanced Technologies of Electromagnetic Field Assisted Consolidation of Materials», held in Moscow, Russian Federation on 29-31 August 2016. This workshop is regularly organized by *Key Laboratory for Electromagnetic Field Assisted Materials Processing* of *National Research Nuclear University MEPhI* (*Moscow Engineering Physics Institute*), Russian Federation.

The Scientific Workshop was sponsored by the Competitiveness Growth Program of MEPhI ("5–100" Russian Academic Excellence Project - http://5top100.ru).

One of the main goals of this Scientific Workshop was to train young scientists in matters of advanced technologies of electromagnetic field assisted consolidation of powder materials. Meetings were held in the form of review lectures of the leading scientists with free discussion and personal oral and poster presentations of the participants.

The general scope of the Scientific Workshop covered the advanced technologies of electromagnetic field assisted consolidation of materials.

The workshop was attended by 67 participants from 5 countries (Russia, Belarus, Ukraine, Latvia and USA). During the three days of the workshop 14 invited lectures, 35 oral presentations were delivered. However, only some papers were qualified for being published in the Open Access IOP Conference Series: Materials Science and Engineering (MSE).

Based on the peer-review of reports presented at the workshop 22 selected papers have been included in this volume. These papers present new research in the various fields of technology of electromagnetic field assisted consolidation of powder materials. We sincerely hope that the reports presented in this volume will contribute to the advancement of knowledge not only in these fields, but also others.

The organizing committee of the workshop thanks all the participants for their fruitful work and personal contribution to the development of these conference proceedings.

The Editors

Dr. Eugene A. Olevsky
Dr. Evgeny G. Grigoryev

LISTS OF COMMITTEES

International advisory committee

Chair: Olevsky Eugene A. – San Diego State University (USA)

Alymov Mikhail I. - Institute of Structural Macrokinetics and Materials Science RAS (Russia)
Belyavin KlementiyE. – Belarusian National Technical University (Belarus)
Glushchenkov Vladimir A. – Samara National Research University (Russia)
Grigoryev Evgeny G. – NRNU «MEPhI» (Russia)
Dudina Dina V. – Institute of Solid State Chemistry and Mechanochemistry of the Siberian Branch of the Russian Academy of Sciences (Russia)
Khasanov Oleg L. – Tomsk Polytechnic University (Russia)
Kuznetchik Oleg O. – The State Scientific Institution «Powder Metallurgy Institute» (Belarus)
Minko Dmitriy V. – Belarusian National Technical University (Belarus)
Mironov Viktor A. – Riga Technical University (Latvia)
Sizonenko Olga N. – Institute of Impulse Processes and Technologies NAS (Ukraine)
Stolyarov Vladimir V. –Mechanical Engineering Research Institute RAS (Russia)
Strizhakov Evgeny L. – Don State Technical University (Russia)

Technical program committee

Chair: Grigoryev Evgeny G. - NRNU «MEPhI» (Russia)

Guseva Lubov Y. - NRNU «MEPhI» (Russia)
Nefedova Evgenia V. - NRNU «MEPhI» (Russia)
Rubinkovskiy Nikita A. - NRNU «MEPhI» (Russia)
Terekhov Grigoriy P. - NRNU «MEPhI» (Russia)
Yudin Artem V.- NRNU «MEPhI» (Russia)
Zholnin Anatoliy G. - NRNU «MEPhI» (Russia)

Publication committee

Chair: Grigoryev Evgeny G. - NRNU «MEPhI» (Russia)

Nefedova . Evgenia V. - NRNU «MEPhI» (Russia)
Rubinkovskiy Nikita A. - NRNU «MEPhI» (Russia)
Guseva Lubov Y. - NRNU «MEPhI» (Russia)

Peer review statement

All papers published in this volume of *IOP Conference Series: Materials Science and Engineering* have been peer reviewed through processes administered by the proceedings Editors. Reviews were conducted by expert referees to the professional and scientific standards expected of a proceedings journal published by IOP Publishing.

IOP Conf. Series: Materials Science and Engineering **218** (2017) 012001 doi:10.1088/1757-899X/218/1/012001

Effect of specific absorbed power on microwave sintering of 3YSZ ceramics

Yu V Bykov[1], S V Egorov[1], A G Eremeev[1], I V Plotnikov[1], K I Rybakov[1,2], A A Sorokin[1] and V V Kholoptsev[1]

[1] Institute of Applied Physics, Russian Academy of Sciences, Ul'yanov St. 46, Nizhny Novgorod, 603950, Russia
[2] Lobachevsky State University of Nizhny Novgorod, Gagarin Ave. 23, Nizhny Novgorod, 603950 Russia

Abstract. Samples of 3 % yttria-stabilized zirconia (3YSZ) ceramics have been sintered to near full density with no appreciable grain growth using an ultra-rapid microwave sintering process. The sintering experiments were carried out on a 24 GHz / 6 kW gyrotron system for microwave processing of materials with automatic process control. By varying the properties of the thermal insulation surrounding the samples it was possible to vary the microwave power required for heating. The final relative density of 3YSZ ceramic samples microwave heated at a rate of 50 °C/min to a temperature of 1400 °C without isothermal hold varied from 91.6 % when the specific absorbed microwave power was 4 W/cm^3 to 99.4 % when the specific absorbed microwave power was 90 W/cm^3. The specific absorbed power is therefore demonstrated to be the key parameter determining the achievable density in ultra-rapid field-assisted sintering processes.

1. Introduction

The effect of very rapid, so called flash sintering occurring when the sample is subjected to a dc or ac electric field, has recently attracted significant research interest. Authors of most recent studies associate the flash sintering effect with the development of an overheating instability in the materials [1–3]. This instability, commonly known as thermal runaway, is well known in the experiments on microwave heating of dielectric materials – in particular, on microwave sintering. A model explaining the growth of the instability was developed in [4–6]. The instability is observed during the heating of materials in which the effective electric conductivity grows with temperature, which creates a nonlinear positive feedback. It develops when the ratio between the specific power of microwave radiation absorbed per unit volume of material and the specific power of heat losses from its surface exceeds a certain threshold. A similar effect is apparently observed when the role of the volumetric power source is played by the Joule losses of the electric current flowing through the sample, as in the dc/ac flash sintering experiments. If the conductivity vs. temperature dependency and the parameters of the cooling process are known quantitatively, it is possible to calculate the parameters of the thermal instability. Such calculations were accomplished to determine the instability onset for the dc field-assisted sintering of ceramic compacts of various compositions: 3YSZ [2], 3YSZ and $BaTiO_3$ [7], ZnO [3], TiO_2 [8], Al_2O_3 [9], $SrTi_{1-x}Fe_xO_{3-\delta}$ [10]. In the flash sintering experiments the measurable process parameters are the applied voltage, current, furnace temperature and the temperature at the sample surface. However, it is clear that the development of the thermal instability is determined not by separate electric parameters but rather by the relationship between the energetic characteristics of the heating process, viz. specific power of volumetric heat deposition and heat removal. An analysis of many experimental results on flash sintering [11] suggests that regardless of the type of electrical conduction in the material, the transition to rapid nonlinear increase in the electric conductivity of the materials occurs in a narrow range of specific power, 10...50 W/cm^3, while the values of the temperature and electric field strength vary widely (300...1300 °C and 10...1000 V/cm, respectively).

It follows from estimates [12] that the increase in temperature in the bulk of the samples during the development of the thermal instability may reach 800 °C relative to the temperature before the onset of sintering. However, according to [12] such an increase is insufficient to explain the observed ultra-rapid sintering by enhanced rates of thermally activated processes of solid-state diffusion mass transport. The exponential growth of the conductivity *per se* does not explain the accelerated sintering. Yet, the results of the flash sintering studies evidence that ultra-rapid densification is only observed when the specific power of Joule heating exceeds a certain threshold value - that is, when the thermal instability develops.

Recently, it has been shown that ultra-rapid or flash sintering is also observed during microwave sintering of ceramic materials such as Al_2O_3, Y_2O_3, $MgAl_2O_4$, and $Yb(LaY)_2O_3$ [13, 14]. It follows from these studies that a sharp acceleration of sintering occurs due to softening or melting of grain boundaries when the level of specific absorbed microwave power corresponds to the onset of thermal instability. In this paper the effect of the specific absorbed power is investigated in the process of microwave sintering of 3YSZ ceramics.

2. Experimental

Commercially available 3 % yttria-stabilized zirconia powder (TZ-3Y, Tosoh corporation, Japan) was used in the present work. According to the producer's certificate, the powder has the following physical and chemical properties: crystallite size – 27 nm, particle size (D_{50}) – 0.6 μm, specific surface area – 16 ± 4 m^2/g; impurity content (wt. %): $Al_2O_3 \leq 0.1$, $SiO_2 \leq 0.02$, $Fe_2O_3 \leq 0.01$, $Na_2O \leq 0.04$. The powder was used to fabricate plates by slip casting. The plates were dried for several days in air, then calcined for 2 hours at 800 °C in air. The samples in the form of discs, 12 mm in diameter and approximately 2.5 mm in height, were cut out from the plates. The green density of the samples was 49.5 % of the theoretical density. Two dimples for positioning thermocouple heads, 1 mm in diameter and 1.25 mm in depth, were drilled on the surface of the samples in the centre and 1 mm away from the edge.

The samples were heated in the applicator of a gyrotron system with a microwave power up to 6 kW at a frequency of 24 GHz equipped with a computerized feedback power control circuit [15]. The powder compacts were sintered in air at normal pressure. To limit heat losses from the samples undergoing microwave heating, two types of thermal insulation arrangements were used. In one case, the samples were placed in the centre of a cylindrical quartz crucible, 100 mm in diameter and 100 mm in height, which was filled with coarse alumina powder. In the other case, the samples were placed in a container fabricated from plates of highly porous, low microwave absorbing material Valox-1750 (Thermoceramica Ltd., Russia).

The samples were heated to a preset maximum sintering temperature chosen from the range 1100 – 1500 °C at a preset fixed heating rate (10, 15, 50, 100, 150 °C/min). The temperature was measured using B-type thermocouples. The process control was based on the reading of the thermocouple located in the centre of the sample. The heating was terminated when the preset maximum temperature was reached, i.e., sintering with zero hold time was performed. The microwave power was switched off automatically by the computer control system and the sample cooled down along with the thermal insulation surrounding it. For comparative study of densification the samples were heated conventionally in identical regimes using the resistive furnace "Thermoceramica-1700" (Thermoceramica Ltd., Russia).

The density of the sintered samples was determined by Archimedes weighing in distilled water with an estimated accuracy of ± 0.01 g/cm^3. The microstructure of the sintered samples was studied by scanning electron microscopy (JEOL JSM-6390 LV). The phase composition of the sintered samples was analyzed using a Rigaku Ultima IV X-ray difractometer.

IOP Conf. Series: Materials Science and Engineering **218** (2017) 012001 doi:10.1088/1757-899X/218/1/012001

3. Results and discussion

To analyze the influence of the specific absorbed microwave power on the process of densification of the material, series of heating runs were accomplished in identical regimes differing in the conditions of convective and radiative heat removal from the surface of the samples. The specific power spent on the heating of the samples, P_v, was determined from the energy balance equations, using the values of the heating and the cooling rates recorded immediately before and after the time instant when the maximum temperature was achieved and the microwave power was switched off [13, 14]:

$$P_v = \rho C\left(\frac{dT}{dt}\right)_+ + P_{hl} \qquad P_{hl} = \rho C\left|\left(\frac{dT}{dt}\right)_-\right|, \tag{1}$$

where ρ is density, C is specific heat capacity of the material of the sample, P_{hl} is the power in heat losses, $(dT/dt)_+$ and $|(dT/dt)_-|$ are the rates of heating and cooling before and after the microwave power switchoff, respectively. The value of P_v can be varied in experiments by either varying the heating rate (that depends on the input microwave power) or the cooling rate (that depends on the heat removal / thermal insulation conditions).

The variation of the heat removal conditions was implemented in this study by using two types of thermal insulation arrangements with considerably different thermophysical and electrophysical properties. It should be noted that within the approach adopted in this paper there is no need in knowing these properties in detail. The power in heat losses, P_{hl}, which is needed to derive the specific absorbed power, is characterized by measuring the cooling rate. This makes this method especially attractive from the viewpoint of practice. In fact, the absence of information about the microwave absorption coefficients of the thermoinsulating materials and their temperature dependencies, as well as the lack of data on their thermophysical properties make it not possible to calculate or estimate the heat removal characteristics. Moreover, even if such data were available, the calculations would be not much reliable because the 3D temperature fields in the objects undergoing heating are not known.

To illustrate the different properties of two thermal insulation arrangements, shown in figure 1 are the temperature data recorded during heating of the samples at a rate of 50 °C/min to a temperature of 1400 °C and subsequent cooling using the Al_2O_3 powder filling (a) and Valox-1750 plates (b) as thermal insulation. In the former case the cooling rate at the beginning of the cooling process was 70 °C/min, and in the latter case – 1360 °C/min.

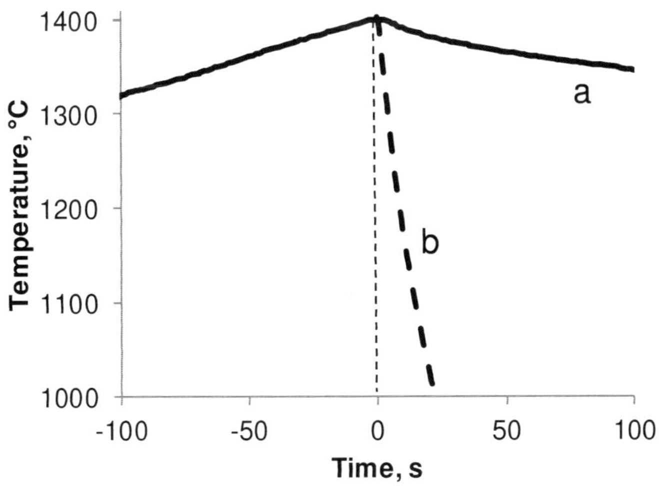

Figure 1. Temperature vs. time during microwave heating of 3YSZ samples at a rate of 50 °C/min to 1400 °C and subsequent cooling using the Al_2O_3 powder filling (a) and Valox-1750 plates (b) as thermal insulation. The zero time instant corresponds to the microwave power shutoff at maximum temperature.

The difference in the cooling rates under different thermal insulation conditions is associated with the difference in the temperature distributions formed due to the absorption of microwave radiation. In the case of Al_2O_3 powder filling, the microwave absorption in the powder layer with a thickness of about 4 cm results in its considerable heating. The measurements of the temperature difference between the centre of the sample and the point located 1 mm away from its edge give an idea of the temperature

IOP Conf. Series: Materials Science and Engineering **218** (2017) 012001 doi:10.1088/1757-899X/218/1/012001

value at the side of the thermal insulation arrangement facing the sample. When the temperature of the sample is 1300...1500 °C, the temperature difference on the radial distance of 5 mm is 20 – 25 °C in the case of Al_2O_3 powder filling and about 100 °C in the case of Valox-1750 plates used as thermal insulation. In the latter case, the temperature gradients arising in the samples reach about 200 °C/cm, which sometimes resulted in partial cracking of the samples.

Listed in table 1 are the values of the specific absorbed power – estimated according to equations (1) and averaged over several experimental runs – for the processes of heating to temperatures 1100...1500 °C using different thermal insulation arrangements. For estimates it was taken that $\rho \approx 6$ g/cm^3 and $C \approx$ 0.7 J/g·°C.

Table 1. Specific absorbed microwave power vs. heating rate for two different thermal insulation arrangements.

Heating rate, °C/min	Specific absorbed microwave power, W/cm^3	
	Thermal insulation: Al_2O_3 powder ($1300\ °C \le T \le 1500\ °C$)	Thermal insulation: Valox-1750 plates ($1100\ °C \le T \le 1400\ °C$)
15	4	60
50	8	90
100	12	120

In the case of "less efficient" thermal insulation (Valox-1750 plates) the values of the specific absorbed power are more than an order of magnitude higher. It occurs that this results in considerably higher final densities of the sintered samples. For example, the density of 95.5 % achieved using the "less efficient" thermal insulation at a temperature of 1100 °C (P_v = 60 W/cm^3) is only obtained at a temperature of 1450 °C when the "more efficient" thermal insulation (Al_2O_3 powder filling) is used and the specific absorbed power is much lower (P_v = 4 W/cm^3).

Shown in figure 2 are the dependencies of the relative density of the samples, sintered at different temperatures, on the specific absorbed power during heating. When the samples were microwave heated in the "more efficient" thermal insulation their relative density varied from 55.3 to 96.2 % of the theoretical value at variation of temperature from 1100 °C to 1400 °C. When the "less efficient" thermal insulation was used, the relative density varied from 95.5 to 99.4 % at the same variation of temperature. It should be noted here that 3YSZ ceramic samples with densities exceeding 99 % have been fabricated by ultra-rapid microwave sintering in the regimes without the isothermal heating stage at maximum temperature.

It can be argued that the levels of the specific absorbed power that are characteristic of the processes using the "more efficient" thermal insulation (2 W/cm$^3 \le P_v \le$ 12 W/cm^3) are insufficient to trigger the thermal instability, whereas the levels of the specific absorbed power in the processes using the "less efficient" thermal insulation (60 W/cm$^3 \le P_v \le$ 120 W/cm^3) do lead to the development of the instability. It occurs that the presence of the instability is a necessary condition of obtaining high final density in an ultra-rapid sintering process. This observation is consistent with the results of analysis of dc/ac field-assisted flash sintering experiments [11] which indicate that the development of thermal instability in various materials begins at specific absorbed power levels on the order of 10...50 W/cm^3 and results in ultra-rapid ("flash") densification.

Shown in figure 3 are SEM images of the polished surfaces of the samples sintered under microwave heating at a rate of 50 °C/min to a temperature of 1400 °C in the "more efficient" thermal insulation (figure 3a, P_v = 12 W/cm^3) and "less efficient" thermal insulation (figure 3b, P_v = 120 W/cm^3). The average grain size in both samples is close to the initial grain size (D_{50} = 0.6 μm). This demonstrates that near full density 3YSZ ceramics (ρ = 99.4 % ρ_{th}) with practically no grain growth can be obtained by ultra-rapid microwave sintering without isothermal hold.

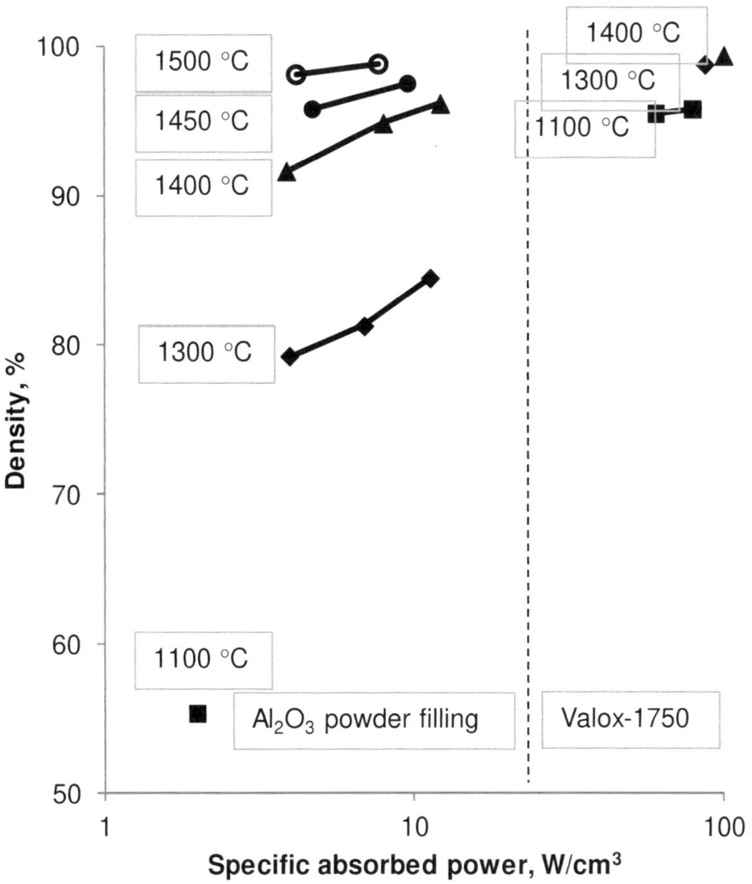

Figure 2. Relative density of the sintered 3YSZ samples vs. specific absorbed microwave power for different maximum temperatures and two types of thermal insulation – Al_2O_3 powder filling or Valox-1750 plates.

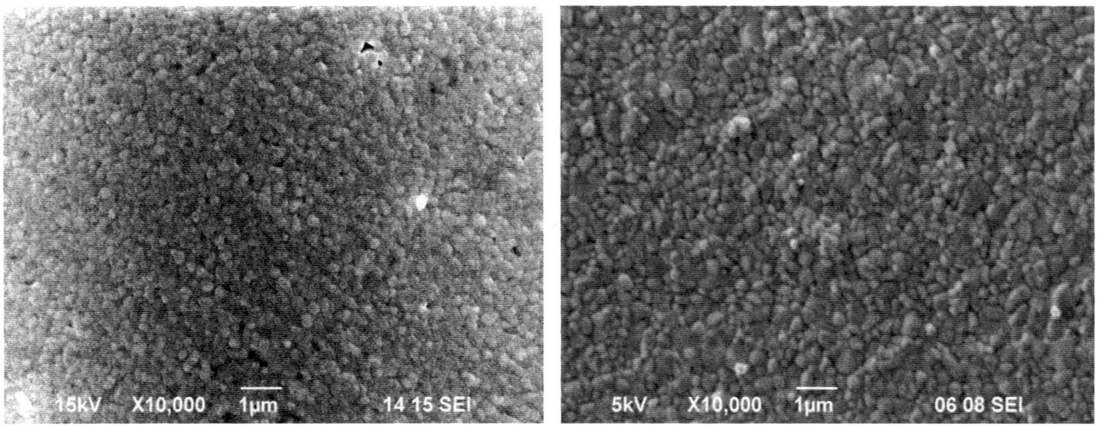

(a) (b)

Figure 3. SEM images of the polished surfaces of the samples sintered under microwave heating at a rate of 50 °C/min to a temperature of 1400 °C: (a) in the "more efficient" thermal insulation (Al_2O_3 powder filling, P_v = 12 W/cm³); (b) in the "less efficient" thermal insulation (Valox-1750 plates, P_v = 120 W/cm³).

4. Conclusion

Near full density 3YSZ ceramics have been obtained by ultra-rapid microwave sintering without isothermal hold. Along with temperature, a critical parameter that characterizes the ultra-rapid sintering is the specific microwave power absorbed in the sample per unit volume. At a temperature of 1400 °C and the specific absorbed power about 90 W/cm^3 the achieved density of the sintered samples is 99.4 % of the theoretical value. Due to high heating rates and the absence of isothermal hold the average grain size in the sintered samples is practically equal to the particle size of the initial powder material (0.6 μm).

Acknowledgment

This research was supported in part by Russian Foundation for Basic Research, grant # 16-08-00736.

References

[1] Grasso S, Sakka Y, Renndtorff N, Hu C, Maizza G, Borodianska H and Vasilkiv O 2011 *J. Ceram. Soc. Japan* **119** 144
[2] Todd R I, Zapata-Solvas E, Bonilla R S, Sneddon T and Wilshaw P R 2015 *J. Eur. Ceram. Soc.* **35** 1865
[3] Zhang Y, Jung J-I and Luo J 2015 *Acta Mater.* **94** 87
[4] Roussy G, Bennani A and Thiebaut J-M 1987 *J. Appl. Phys.* **62** 1167
[5] Kriegsmann G A 1992 *J. Appl. Phys.* **71** 1960
[6] Kriegsmann G A 1991 Microwave Heating of Ceramics *Microwaves: Theory and Applications in Materials Processing (Ceramic Transactions vol 21)* ed D E Clark, F D Gac *et al.* (Cincinatti, OH: The American Ceramic Society) pp 177-183
[7] Pereira da Silva J G, Al-Qureshi H A, Keil F and Janssen R 2016 *J. Eur. Ceram. Soc.* **36** 1261
[8] Zhang Y, Nie J and Luo J 2016 *J. Ceram. Soc. Japan* **124** 296
[9] Biesuz M and Sglavo V M 2016 *J. Eur. Ceram. Soc.* **36** 2535
[10] Shomrat N, Baltianski S, Dor E, Tsur Y 2017 *J. Eur. Ceram. Soc.* **37** 179
[11] Raj R 2016 *J. Am. Ceram. Soc.* DOI: 10.1111/jace.14178
[12] Raj R 2012 *J. Eur. Ceram. Soc.* **32** 2293
[13] Bykov Yu V, Egorov S V, Eremeev A G, Kholoptsev V V, Rybakov K I and Sorokin A A 2015 *J. Am. Ceram. Soc.* **96** 3518
[14] Bykov Yu V, Egorov S V, Eremeev A G, Kholoptsev V V, Plotnikov I V, Rybakov K I and Sorokin A A 2016 *Materials* **9** 684
[15] Bykov Yu, Eremeev A, Glyavin M, Kholoptsev V, Luchinin A, Plotnikov I, Denisov G, Bogdashev A, Kalynova G, Semenov V and Zharova N 2004 *IEEE Trans. Plasma Sci.* **32** 67

IOP Conf. Series: Materials Science and Engineering **218** (2017) 012002 doi:10.1088/1757-899X/218/1/012002

Spark plasma sintering of high-strength lightweight ceramics

V N Chuvil'deev, M S Boldin, A A Popov and A V Nokhrin

Lobachevsky State University of Nizhny Novgorod, National Research University Research and Development Institute of Physics and Technology, 23 Prospect Gagarina (Gagarin Avenue), Nizhny Novgorod, 603950, Russia.

E-mail: boldin@nifti.unn.ru

Abstract. This research paper provides an illustration of how to use the Spark Plasma Sintering technology (SPS) for powder materials in order to obtain lightweight ceramics Al_2O_3/ZrO_2 with enhanced strength properties. Optimization of SPS modes helps to produce ceramics with grain size of less than 400 nm, microhardness $H_V = 24$ GPa, and crack resistance $K_{IC} = 4.2$ MPa·m1/2.

1. Introduction

Alumina is one of the most advanced ceramic materials for a wide range of structural applications, including aerospace engineering, owing to a combination of high hardness, thermal and chemical stability on the one hand, and availability on the other hand. One of the means of improving alumina performance properties is to develop composite structures by adding into original Al_2O_3 ceramic such fine powders as ZrO_2, SiC, TiC, TiN, etc. [1-5].

Due to high rupture resistance, Al_2O_3/ZrO_2 system has gained widespread use in industry [6]. Tetragonal-monoclinic phase transition of zirconium dioxide under high mechanical stress ensures high crack resistance of this composite material when subjected to loadings: a zirconium dioxide particle undergoes structural changes when appears on a crack path. During phase transition, ZrO_2 expands ensuring stress relaxation at the crack tip, thus further crack propagation slows down [7].

Traditional Al_2O_3/ZrO_2 composites contain from 5 volume % to 50 volume % of ZrO_2 [6, 8-9]. Since the specific weight of zirconia exceeds the specific weight of the alumina ~ 1.4 times, the weight of a finished product with 25 volume % of ZrO_2 is 10% bigger compared to the weight of a product made of pure Al_2O_3.

In this regard, ceramic materials based on alumina with low content of zirconia are becoming ever more relevant for aerospace applications, as they are lighter but ensure strength properties that are as good as those of traditional Al_2O_3/ZrO_2 composites.

In recent decades, the high-speed compaction technology known as Spark Plasma Sintering (SPS) has become a frequent practice in powder engineering [10]. One of the key features of SPS technology that distinguishes it from other consolidation technologies is its high control precision of sintering parameters. Review [11] shows that optimization of SPS modes allows to obtain almost any oxide ceramics with density close to 100% and grain growth factor below 2. At the same time, ultrafine-grained structure of sinter materials ensures high strength properties.

Nano powders, surfactants, and ultrasonic processing ensure very even distribution of strengthening dispersed particles in the base material and therefore significantly reduce the volume fraction of ZrO2 particles required to boost crack propagation resistance.

IOP Conf. Series: Materials Science and Engineering **218** (2017) 012002 doi:10.1088/1757-899X/218/1/012002

From this perspective, it looks very promising to combine innovative approaches to obtaining powder composite systems and SPS technology to produce Al_2O_3-based lightweight high-strength ceramics, which is the goal of the present work.

2. Experimental

Al_2O_3 powders made by "Taimei Chemicals Co., Ltd." (Japan) and ZrO_2 made by "Pangea Int., Ltd" (China) constituted the target of this research. The microstructure of original powders is shown on figure 1.

Al_2O_3/ZrO_2 composite mixtures were obtained through mixing original powders for 4 hours in a planetary mill "FRITSCH - Pulverisette 6" with grinding media of zirconium dioxide in isopropyl alcohol.

Samples 12 mm in diameter and 3 mm high was carried out using spark plasma sintering method in a sintering unit "DR. SINTER model SPS-625 Spark Plasma Sintering System" (SPS SYNTEX INC. Ltd., Japan). The temperature was measured using a pyrometer centered on the outer surface of the graphite mold. The heating rate was 50 ^{0}C/min, while the applied load remained under 70 MPa. Sintering took place in 6 Pa vacuum. The sintering temperature did not exceed 1200 ^{0}C. Sintering modes are shown in table 1.

The density of sintered samples was measured using the hydrostatic weighing in distilled water using Sartorius CPA scales. Accuracy of density calculations was ± 0.005 g/cm^3.

Vickers hardness (H_V) was measured using an automated microhardness tester "Struers Duramin-5" with a 2 kg load. Crack resistance K_{IC} was calculated using the Palmqvist method. Measuring accuracy of H_V and K_{IC} was ± 1.5 GPa and ± 0.5 MPa·m$^{1/2}$ respectfully.

The microstructure of samples was studied with the help of a scanning electron microscope Jeol JSM-6490. The average grain size in sintered ceramics was calculated after analyzing at least 400 grains. The microstructure of the obtained samples is shown on figure 2.

Figures 2a-2e show that new approaches to powder mixing help to improve evenness of distribution of ZrO2 dispersed particles in sintered ceramics.

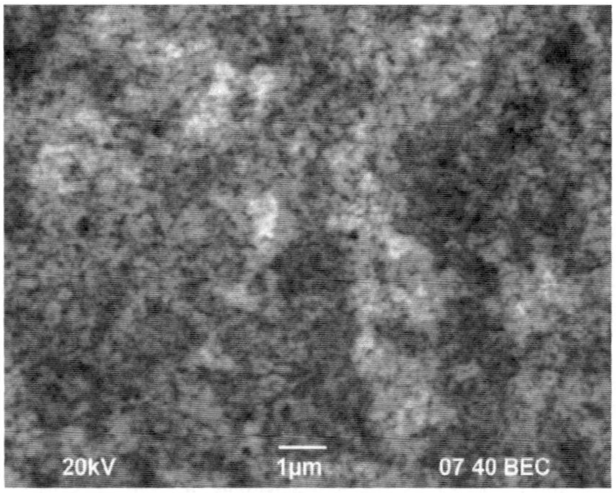

(a)

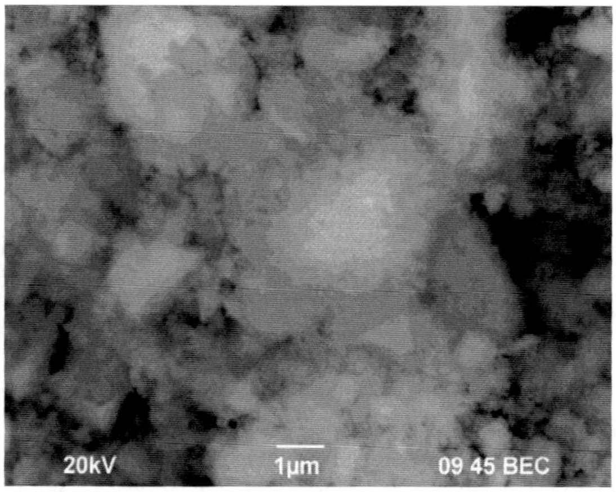

(b)

Figure 1. Microstructure of starting powders: (a) Al_2O_3 powder (d ~ 200 nm), (b) ZrO_2 powder (d ~ 30 nm). Scanning electron microscopy (SEM).

Table 1. Properties of composite ceramics obtained through SPS method.

Material	T_{sint} (^{0}C)	t (min)	d (μm)	ρ_{relat} (%)	H_V (GPa)	K_{IC} (MPa·m$^{1/2}$)
Al_2O_3	1200	0	0.5	99.5	22.3	2.2
Al_2O_3 + 1.5 vol.% ZrO_2	1200	0	0.3	96.1	20.9	3.2
	1200	3	0.4	99.6	24.0	4.2
	1200	10	0.6	99.6	22.0	2.7
	1200	30	0.9	99.6	21.0	2.8
Commercial ceramics (hot pressure sintering) Al_2O_3+5 vol.% ZrO_2	n.d.	n.d.	1.5	99.1	19.1	3.0

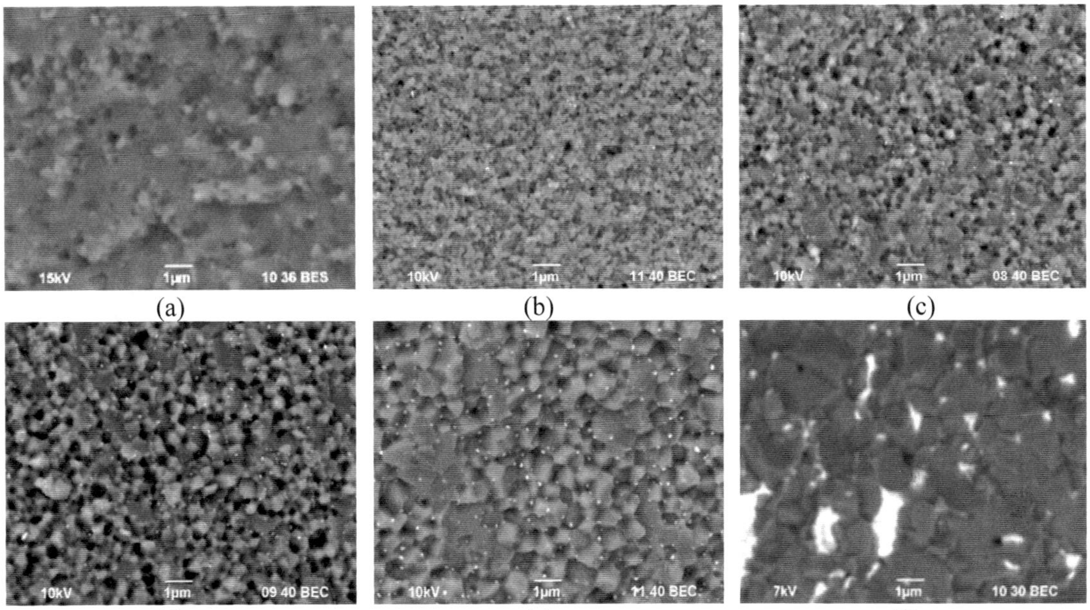

Figure 2. Microstructure of ceramics: (a) Al_2O_3, (b) Al_2O_3/ZrO_2 (at t = 0 min), (c) Al_2O_3/ZrO_2 (at t = 3 min), (d) Al_2O_3/ZrO_2 (at t = 10 min), (e) Al_2O_3/ZrO_2 (at t = 30 min), (f) commercial ceramics Al_2O_3+5 volume % ZrO_2. Bright particles – ZrO_2, dark particles – Al_2O_3. SEM

3. Results and discussion

It is well-known that the strength of sintered ceramic materials depends on their density and grain size. The larger is the grain, the lower is the strength. Besides, coarsening of grain causes increase in diffusion paths, which impedes higher density during sintering and consequently has a detrimental effect on strength.

The size of grain in sintered ceramics depends on the initial size of powder particles, size and volume fraction of dispersed second phase particles and pores. It is crucial to note that unless a continuous net of grain boundaries is formed, coarsening of grain is impeded. A net of grain boundaries is considered to be formed at relative density of $\rho \sim 90\%$. During sintering of spherical particle with r radius, the sample microstructure at that moment consists of equiaxed grains, with grain size $d_{grain} = r$ and pores size $d_{pore} = 0.15 \cdot r$. From this point an intense grain growth is observed [12].

In this context, in order to obtain materials with maximum density and minimum grain size it is required to optimize temperature and isothermal time after the material passes through $\rho \sim 90\%$.

At high sintering temperature, boundary migration speed is generally high, boundaries may "come off the pores" leaving them in the grain volume. In this case, pores dissolution rate as well as contraction will be controlled by volume diffusion rather than grain boundary diffusion. Since [13] the volume diffusion coefficient at Al_2O_3 sintering temperatures (1200 ^{0}C) is two orders less than the grain boundary diffusion coefficient, sintering-out of voluminous pores is rather slow and occurs in modes when boundaries come off pores. Dense materials are therefore hard to obtain.

To eliminate the come-off effect, boundary migration speed shall be slowed down. It can be achieved through reducing the sintering temperature after reaching the 90%-density point. This approach is known as "Two Step Sintering», «Multi Step Sintering», «Rate Control Sintering» [14-16].

Lowering the temperature at the second sintering stage helps to reduce grain growth rate while maintaining pores dissolution rate at a reasonable level. This becomes possible through SPS technology that offers ample opportunities for temperature control.

4. Conclusion

Due to high accuracy achieved when controlling sintering parameters, SPS technology enables multi-step sintering modes and high-density nano- and ultrafine-grained materials with improved physical and mechanical properties, which opens up new opportunities to optimize the existing ceramics and develop new structural ceramics.

Acknowledgements

The authors wish to thank the Russian Foundation for Basic Research (grant No. 15-48-02585 r_povolzhye_a), the President's Council (grant No. SP-1775.2015.1).

References

[1] Niihara K 1991 *Journal of The Ceramic Society of Japan* **99 [10]** 974
[2] Sternitzke M 1998 *J. Eur. Ceram. Soc* **17** 1061
[3] Sternitzke M, Derby B and Brook R J 1998 *J. Am. Ceram. Soc* **81 (1)** 41
[4] Young M K, Won T K and Young-Wook K 2004 *Ceram. Int.* **30** 2081
[5] Barry J and Byrne G 2001 *Wear* **247** 139
[6] Chuvil'deev V N, Boldin M S, Dyatlova Ya G, Rumyantsev V I and Ordanyan S S 2015 *Inorganic Materials* **51 (10)** 1047
[7] Fistul V I 1995 *New materials. Status, Problems and Prospects* (Moscow: The National University of Science and Technology MISiS Press) p 142
[8] Garshin A P, Gropyanov V M, Zaitsev G P and Semenov S S 1997 *Engineering ceramics* (Saint Petersburg: Saint Petersburg Mining University Press) p 726
[9] Matsumoto Y, Hirota K, Yamaguchi O, Inamura S and Miyamoto H 1993 *J. Am. Ceram. Soc.* **76 (10)** 2677
[10] Munir Z A and Quach D V 2011 *J. Am. Ceram. Soc.* **94 (1)** 1
[11] Orru R, Licheri R, Locci A M, Cincotti A, and Cao G 2009 *Materials Science and Engineering R* **63** 127
[12] Fang Z Z 2010 *Sintering of advanced materials fundamentals and processes* (Woodhead Publishing Limited)
[13] Frost H J and Ashby M F. 1982 *Deformation-mechanism maps: the plasticity and creep of metals and ceramics* (Pergamon Press)
[14] Hesabi Z R, Haghighatzadeh M, Mazaheri M, Galusek D and Sadrnezhaad S K 2009 *J. Eur. Ceram. Soc* **29** 1371
[15] Reddy K M, Kumar N and Basu B 2010 *Scripta Materialia* **62** 435
[16] Ragulya A V, and Skorokhod V V 1998 *Metal Powder Report* **53 (7-8)** 45

IOP Conf. Series: Materials Science and Engineering **218** (2017) 012003 doi:10.1088/1757-899X/218/1/012003

Fast synthesis and consolidation of porous FeAl by pressureless Spark Plasma Sintering

D V Dudina[1,2,3], A E Brester[2,3], A G Anisimov[1], B B Bokhonov[2,4], M A Legan[1,3], A N Novoselov[1], I N Skovorodin[5] and N F Uvarov[2,3]

[1]Lavrentyev Institute of Hydrodynamics SB RAS, Lavrentyev Ave. 15, Novosibirsk, 630090, Russia
[2]Institute of Solid State Chemistry and Mechanochemistry SB RAS, Kutateladze str. 18, Novosibirsk, 630128, Russia
[3]Novosibirsk State Technical University, K. Marx Ave, 20, Novosibirsk, 630073, Russia
[4]Novosibirsk State University, Pirogova str. 2, Novosibirsk, 630090, Russia
[5]Institute of Automation and Electrometry SB RAS, Koptyug Ave. 1, Novosibirsk, 630090, Russia

E-mail: dina1807@gmail.com

Abstract. We report one-step fast synthesis and consolidation of iron aluminide FeAl of high open porosity by pressureless reactive Spark Plasma Sintering (SPS). The starting material of the Fe-40at.%Al composition was a mixture of an iron powder with an average particle diameter of 4 μm and an aluminum powder with an average particle diameter of 6 μm. The rationale behind the choice of the SPS as a processing technique and fine and comparable sizes of the two reactants for the synthesis of high-open porosity FeAl was realization of fast full chemical conversion of Fe and Al into single-phase FeAl reducing the time available for the compact shrinkage. According to the XRD phase analysis, single-phase FeAl compacts formed after SPS at 800 and 900°C. These compacts had open porosities of 41 and 46%, respectively. The transverse rupture strength of the compacts sintered at 700-900°C was found to change little with the sintering temperature in the selected range.

1. Introduction

Porous iron aluminides FeAl and Fe_3Al are promising materials for making high-temperature gas filters due to their high corrosion resistance in sulfur-containing atmospheres [1]. Reactive sintering presents an efficient fabrication method of porous iron aluminides. Gao *et al* [2-3] used this approach and obtained porous FeAl by cold pressing and sintering of Fe+Al mixtures in a vacuum furnace. Relatively coarse powders were used and sintering was conducted at 1200°C for 1 h. The formation mechanism of porous structures during the reaction between iron and aluminum powders remains a matter of active discussion [2-5]. A successful processing method of porous FeAl would be the one that allows achieving fast chemical transformation of the reaction mixture into a single-phase product while maintaining high porosity (limiting shrinkage of the compact).

The advantages of using electric current as a means of heating of the reaction mixture is fast heating and reactivity enhancement relative to conventional processes of heating in a furnace. As was reviewed by Orru *et al* [6], Spark Plasma Sintering (SPS), a sintering technique that uses DC pulses, can be used for the production of both dense and porous materials. In recent years, there has been a growing interest to the possibilities that SPS can offer in the area of porous materials [7-10]. Two main approaches can be used for the synthesis of porous materials by SPS: sintering using space holders [7-9], which are removed from the compact after consolidation, and sintering without the application of pressure [5, 10].

In this work, we report, for the first time, one-step fast synthesis and consolidation of single-phase high-open porosity iron aluminide FeAl by reactive pressureless SPS.

2. Materials and Methods

Carbonyl iron (99%, average size 4 μm, "Syntez-PKZh", Dzerzhinsk, Russia) and gas-atomized aluminum (99.9%, PAD-6, average size 6 μm, "VALKOM-PM", Volgograd, Russia) were used as the reactants. The mixture of the Fe-40at.%Al composition was prepared. The powders were dry-mixed in a horizontal low-energy mixer for 3 h. Spark Plasma Sintering was carried out using a SPS Labox 1575 apparatus (SINTER LAND, Inc., Japan). The temperature during the SPS was controlled by a pyrometer focused on a near-through hole in the die wall. The powder mixtures were sintered into 3-mm thick compacts of 20 mm diameter at 700, 800 and 900°C. The heating rate was 70°C min^{-1} and the holding time at the maximum temperature was 3 min. Very fast synthesis was also attempted using a heating rate of 100 °C min^{-1} and no holding at the maximum temperature. The as-mixed Fe-40at.%Al powders were not pressed before the SPS. The powders were poured into the die and the upper surface of the loosely packed powder sample was rendered flat by a long punch with a diameter equal to that of the punch inserted in the die cavity during the SPS. Then the long punch was extracted from the die and a short punch was inserted. The density of the powder mixture before sintering was 38% (calculated using the theoretical density of the Fe-40at.%Al composition). A schematic of the set-up used for the pressureless SPS experiments is shown in Figure 1. The pressureless SPS experiments had good reproducibility in terms of the resultant sample thickness, porosity and phase composition. Comparative experiments were conducted using a hot press at a heating rate of 50°C min^{-1}.

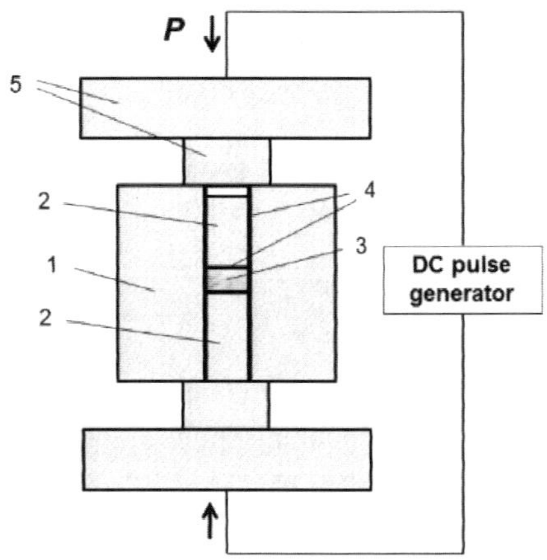

Figure 1. Spark Plasma Sintering set-up for conducting pressureless sintering: 1 – graphite die, 2 – short graphite punches, 3 – sample, 4 – graphite foil, 5 – graphite spacers. The upper punch is in constant contact with the sample during sintering and lowers its position as the sample shrinks. A force of 3 kN was applied to the assembly to establish proper contact conditions between the punches and the spacers for the electric current to flow.

The X-ray diffraction (XRD) patterns were recorded using a D8 ADVANCE diffractometer (Bruker AXS, Germany) with Cu K alpha radiation. The microstructure of the compacts was studied by Scanning Electron Microscopy using a Hitachi-Tabletop TM-1000 microscope (Japan).

The open porosity of the sintered compacts was determined by measuring the weight and calculating the volume of ethanol absorbed by the porous compacts, as described in ref. [11].

The transverse rupture strength (TRS) of the compacts was measured using a method described in refs. [12-13]. A steel ball of 6 mm diameter was used as an indentor to fracture the samples of 20 mm diameter. A universal testing machine Zwick/Roell Z100 (Germany) was used. Three tests were conducted for each sintering temperature. The Poisson's ratio of the porous material was calculated using the value of the Poisson's ratio of the fully dense FeAl and the values of the total porosity of the compacts, as described in ref. [13].

3. Results and Discussion

In our previous investigation [5], it was found that during SPS and hot pressing, the same sequence of reactions takes place in the Fe-40at.%Al mixtures of the micrometer-sized powders:

$$Fe+Al \dashrightarrow Fe+Fe_2Al_5 \dashrightarrow Fe+FeAl \dashrightarrow FeAl.$$

The range of the SPS temperatures used in that work was 500-650°C. Compacts of 10-mm diameter sintered at 650°C were composed of FeAl as a major phase and Fe (residual) as a minor phase. In the present study, the choice of the SPS as a processing technique, sintering temperatures and fine and comparable sizes of the two reactants was dictated by the goal of conducting fast full chemical conversion of Fe and Al into single-phase FeAl reducing the time available for the compact shrinkage.

Figure 2 and Figure 3 show the XRD patterns and fracture surfaces of the 20-mm Fe-40at.%Al compacts sintered 700 and 900°C, respectively. The phase composition of the 20-mm diameter compact sintered at 700°C was similar to that of the 10-mm compact sintered at 650°C [5]. The single-phase FeAl, according to the XRD analysis, formed after the SPS at 800°C. It was also shown that even faster heating (100°C min^{-1}) of the reaction mixture in the SPS up to 900°C without holding at the maximum temperature allows synthesizing single-phase FeAl from the selected reaction mixture. The presence of the $AlFe_3C$ phase in the compacts was due to the presence of carbon in the initial iron powder [5].

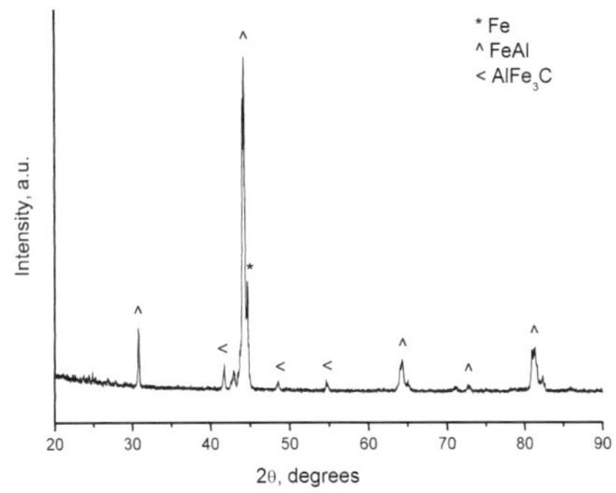

a

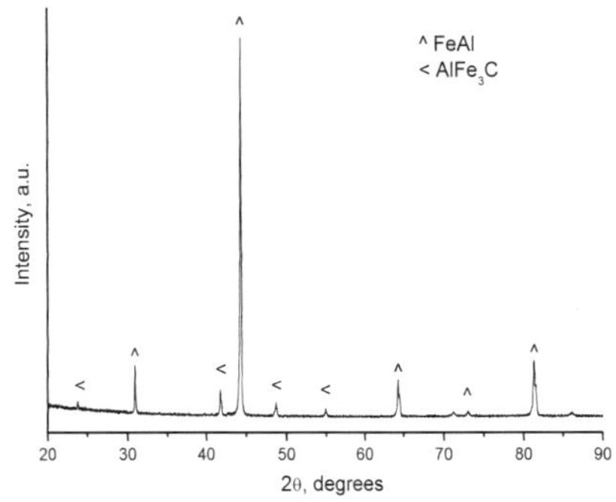

b

Figure 2. XRD patterns of the Fe-40at.%Al compacts obtained by pressureless SPS at 700°C (a) and 900°C (b).

The porous compacts had a uniform structure and did not contain any macrodefects on the sample size scale. Structural differences between the porous materials obtained at 700 and 900°C can be observed. In the former, particles with smooth surfaces - those of the residual iron - can still be distinguished (Fig.3 (a)). The characteristic feature of the latter is the presence of ligaments with a string-like morphology (Fig.3 (b)) indicating a higher degree of chemical transformation of the mixture into the reaction product.

Table 1 summarizes the phase composition, total and open porosity and transverse rupture strength of the FeAl compacts. The content of iron in the compact sintered at 700°C was estimated from the XRD data. The total porosities were calculated based on the precise measurements of the samples' dimensions and using the theoretical density of FeAl (5.67 g cm^{-3}). Karczewski *et al* [14-15] produced FeAl with an open porosity of more than 45-49% using pore forming agents decomposing during sintering and releasing gaseous products. In the present work, high open porosities - 41-46% - were achieved without the use of pore forming agents; therefore, any possible contamination (for example, by carbon) of the porous material caused by decomposition of the pore formers is avoided. A comparative sintering experiment in the hot press at 800°C produced single-phase FeAl with an open porosity of 33%. A lower open porosity can be related to lower heating and cooling rates in this experiment.

The TRS of the compacts sintered at 700-900°C was in the 53-67 MPa range: it did not change significantly with the sintering temperature. Based on these data, a temperature of 800°C can be selected as the processing temperature of porous FeAl by reactive SPS from the selected reactants. At the same time, to reach a slightly higher open porosity and ensure the reaction completeness (some residual iron may still be present in the material sintered at 800 °C in concentrations below the detection limit of the XRD), the synthesis should be conducted at 900°C. The values of TRS of the porous Fe-40at.%Al materials obtained at temperatures below 650°C and still containing the Fe_2Al_5 phase were very low (less than 50% of the TRS of the porous single-phase FeAl).

a b

Figure 3. Fracture surfaces of the Fe-40at.%Al compacts obtained by pressureless SPS at 700°C (a) and 900°C (b).

Table 1. Phase composition (according to the XRD phase analysis), total and open porosities and TRS of FeAl-based porous compacts Spark Plasma Sintered from the Fe-40at.%Al powder mixture.

SPS temperature (°C)	Phase composition	Total porosity (%)	Open porosity (%)	TRS (MPa)	Standard deviation of TRS (MPa)
700	FeAl+5 vol.%Fe	47	41	53.4	5.5
800	FeAl	47	41	67.3	15.9
900	FeAl	53	46	63.0	5.9

4. Conclusions

In this work, we presented the results of the structural characterization and mechanical testing of porous FeAl synthesized and consolidated using reactive pressureless SPS of the Fe-40at.%Al loosely packed powder mixtures. High open porosities of 41-46% are achieved without the addition of pore forming agents using a heating rate of 70°C min^{-1} in the SPS and a holding time at the maximum temperature of 3 min. According to the XRD analysis, single-phase FeAl could be obtained by SPS of the mixtures at 800°C. The porous compacts had a uniform structure and were free from macrodefects. The TRS of the compacts sintered at 700-900°C did not change significantly with the sintering temperature.

The main advantages that are offered by the SPS technique for the synthesis of porous FeAl are the possibilities of high heating rates and relatively high cooling rates of the samples. A limited time of exposure of the reaction mixture to high temperatures allows conducting the synthesis while limiting the shrinkage of the porous material.

References

[1] Henkel D 2011 *Adv. Mater. Proc.* **9** 44
[2] Gao H *et al* 2009 *Intermetallics* **17** 1041
[3] Gao H Y *et al* 2015 *Adv. Powder Technol.* **26** 882
[4] Novák P *et al* 2013 *Intermetallics* **32** 127
[5] Dudina D V *et al* 2016 *Materials* **9** 375
[6] Orru R *et al* 2009 *Mater. Sci. Eng. R* **63** 127
[7] Zhang F *et al* 2010 *Adv. Eng. Mater.* **12** 863
[8] Hakamada M *et al* 2005 *Mater. Trans.* **46** 186
[9] Bokhonov B B *et al* 2015 *J. Phys. Chem. Solids* **76** 192
[10] Cui G *et al* 2016 *Materials* **9** 495
[11] Capek J *et al* 2015 *Mater. & Design* **83** 468
[12] Khaleghi E *et al* 2010 *Scr. Mater.* **63** 577
[13] Wei X *et al* 2015 *Materials* **8** 6043
[14] Karczewski K *et al* 2016 *Mater. Lett.* **164** 32
[15] Karczewski K *et al* 2016 *Mater. Lett.* **178** 268

IOP Conf. Series: Materials Science and Engineering **218** (2017) 012004 doi:10.1088/1757-899X/218/1/012004

Enchasing the mechanical properties of heavy pseudo-alloys of tungsten by high-voltage electric pulse impact

N S Ermakova, S S Bashlykov, E G Grigoriev and V I Golstev

National Research Nuclear University MEPhI, Kashirskoe shosse 31, Moscow, 115409,Russia

E-mail: natalia_ermakova2503@mail.ru

Abstract. The article presents a study of the effect of the high-voltage electric pulse modes of impact under pressure on the structural and mechanical properties of compacts of heavy tungsten alloy VNZh 7-3. Identified an increase in density of material, in microhardness and tensile strength with voltage increase in high-voltage discharge. The optimum exposure mode determinate, which provides the highest level of strength and ductility of the sintered material.

1. Intruduction

Tungsten alloy VNZh 7-3 (90W-7Ni-3Fe) is used in instrumentation engineering, defense industry, to make protective shields and containers for storing radioactive substances, radiological equipment for radiation and contamination monitoring, for production of various kinds of weighting agents, electrical contacts and much more [1, 2]. The aim of this study was to investigate the possibility of the high-voltage sintering of tungsten-based powders and subsequent achievement of high mechanical properties of heavy pseudo-alloys of tungsten.

The principle of high-voltage electric pulse consolidation (HVEPC) is passing of a high-voltage discharge passage (30 kW) with a high current density ($\approx$100 kA$\cdot$sm^{-2}) and short pulse duration (less than 300 μs). This current pulse is obtained by a capacitor bank and passed through a powder filling during the application of external pressure. The rapid increase in temperature at the areas of contact of the powder particles leads to their instantaneous welding. Due to the short sintering time, complete or substantially complete sealing occurs with minimal changes of the microstructure that are undesirable.

As the starting material the powder with composition of 90W-7Ni-3Fe was selected. Average particle size of heavy tungsten alloy VNZh powder was determined by using a laser particle size analyzer. The average particle size was equal 6,03 μm. Figure 1 shows the results of the measurements. Powder density measured with a pycnometer was equal 17.13 g / sm3.

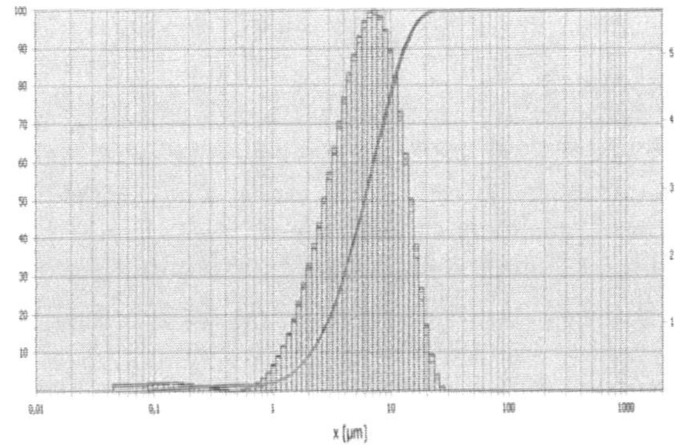

Figure 1. The powder particle size distribution

2. Method of study of structural and mechanical properties of the compacts of heavy tungsten alloy VNZh 7-3 (90W-7Ni-3Fe)

High-voltage electric discharge sintering was conducted in a ceramic matrix with diameter 8 mm for the first batch of samples with cylindrical shape and 10 mm for the second batch of samples with disk shape. The pressure applied to the filling in all the experiments was equal to 45 MPa, the number of pulses was equal to 1. The value of the applied voltage varied from 4.5 kV to 5.8 kV.

After sintering of the powders the microstructure of compacts was studied in a scanning electron microscope, the density was assessed by hydrostatic weighing, and micro-hardness by the Vickers hardness test. Compression testing of cylindrical samples was carried out in accordance with GOST 25.503 standard. The ultimate tensile strength of the material was determined by tensile compression in the center plane of the disk samples by the Brazilian test method [3, 4]. This method applies for assessment of the tensile strength of brittle rock materials. Substantiation of application of this test method to small-sized samples of metallic materials has been provided in the study [5].

3. Experimental results

The first batch of samples was produced in the cylindrical shape with a height to diameter ratio h/d≈3/2. The dependences of density and microhardness of the samples on the charge values are shown in Figures 2 and 3.

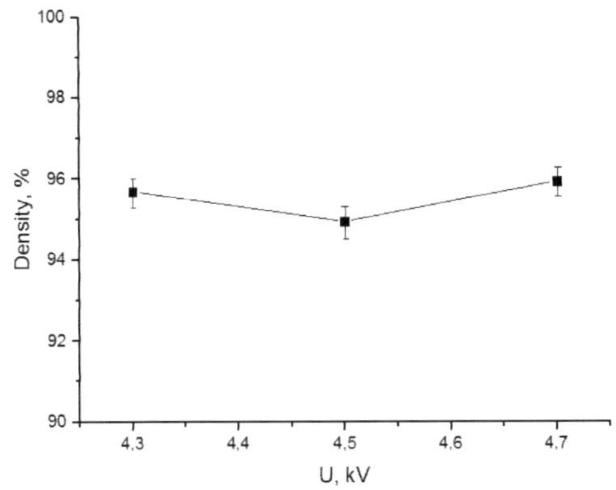

Figure 2. Influence of sintering mode on the density of cylindrical samples

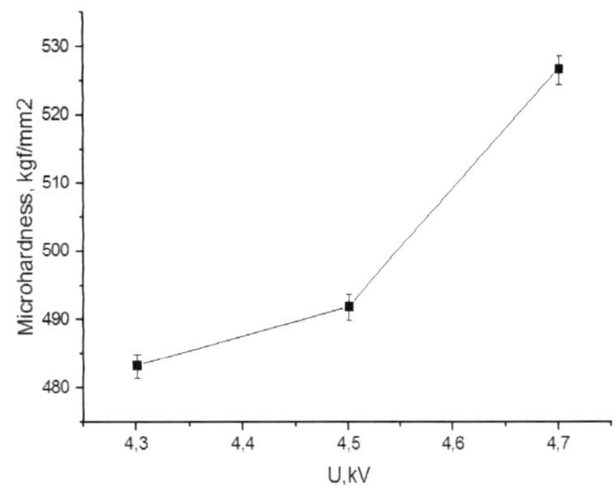

Figure 3. Influence of sintering mode on the microhardness of cylindrical samples

As can be seen from the graphs, the value of the relative density of the sintered compacts is in the same range of values and amounts to 94,9% to 95,9% of theoretical. The microhardness of samples increases with increase of applied voltage.

Compression testing showed that all tested alloys bear compressive stress at room temperature without failure. Figure 4 shows a characteristic diagram of the compressive deformation, which indicates the appearance and development of plastic deformation at the voltage above 1250 MPa. Due to the brittleness of tungsten as a BCC metal, increasing the plasticity of the heavy tungsten alloy is one of the objectives of the current research. Microstructure of the sample obtained by a scanning electron microscope is shown in Fig. 5. The average grain size is approximately 10 microns, i.e. high-voltage sintering contributes to maintaining a fine-grained structure, more uniform distribution of iron-nickel binder and almost total absence of porosity.

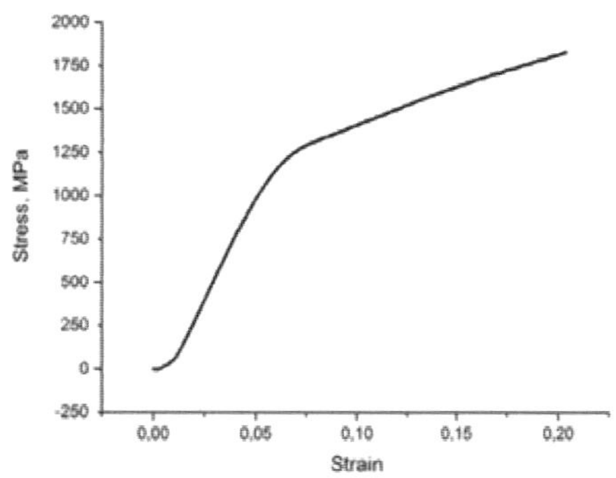

Figure 4. Deformation curve (the voltage equals 5.6 kV)

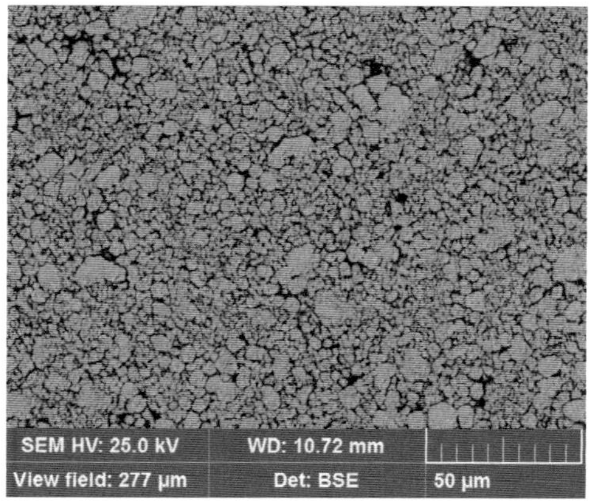

Figure 5. Microstructure of the sample obtained by SEM.

The second batch of samples had a disk shape with a diameter of 10 mm and a thickness of 4.5 to 5.8 mm. Figures 6 and 7 shows the dependences of density and microhardness on the value of pulse voltage. With the increase of the applied voltage an increase in density and microhardness of compacts is observed. The highest values of both characteristics are achieved at the maximum value of discharge 5.8 kV, although stabilized density values obtained at 5.6 kV. Further increase in the microhardness by increasing discharge voltage obviously impractical because it will lead to embrittlement of the material.

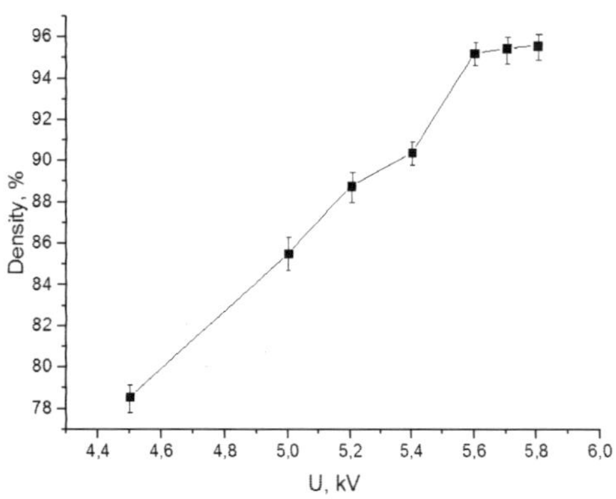

Figure 6. Influence of sintering mode on sample density

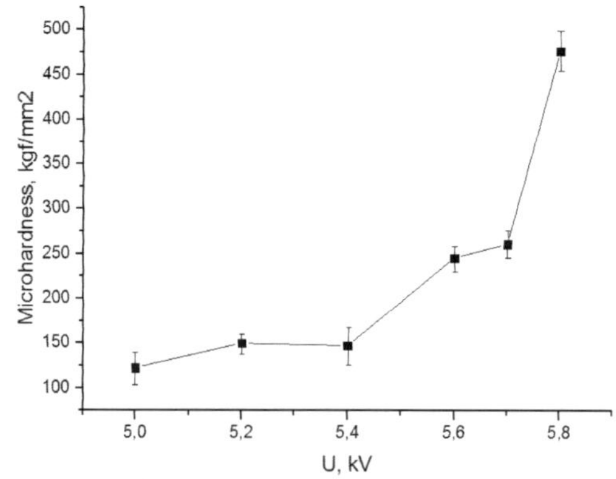

Figure 7. Influence of sintering mode on sample microhardness

Disk compression testing in the center plane by the Brazilian test method was designed to evaluate the tensile resistance of the material under the action of tensile stress and at the same time to identify the plasticity of the material. Figure 8 shows the dependence of the tensile strength of the material on the level of the pulse voltage

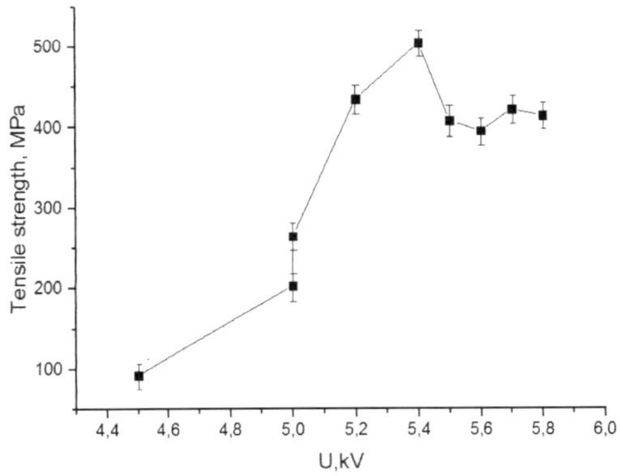

Figure 8. Dependency of the the tensile strength on the applied voltage

a b

Figure 9. Brittle fracture (a) and plastic deformation (b) of the samples

The graph shows an increase of tensile strength of samples at the increasing voltage up a value of 5.4 kV with a subsequent recession. The increase and decrease in strength occurred during the development of plastic deformations, which preceded the brittle failure of the samples. A number of samples did not experienced failure due to a developed plastic flow that can be seen clearly in Fig. 9b. The maximum value of the sample strength was at voltage of 5.4 kV.

4. Conclusions

The studies of compacts obtained from a powder of heavy tungsten alloy VNZh 7-3 (90W-7Ni-3Fe) by the method of high-voltage pulse discharge, revealed the optimum pulse voltage value that results in high structural properties and characteristics of strength and ductility of the material.

References

[1] Povarova K B, Alymov M I and Drozdov A A 2008 Heavy tungsten alloys Heavy tungsten alloys produced from nanopowders *Voprosy Materialovedeniya [Problems of Materials Science]* **2 (54)** 94-99

[2] Afanaseva S A, Belov N N, Biryukov I A et al 2012 Impact properties of tungsten-based alloys at high-speed interaction *Russian Physics Journal* **55 (11)** 35-40

[3] ASTM D3967-95a. Standard Test Method for Splitting Tensile Strength of Intact Rock Core Specimens

[4] Wang Q Z, Jiaa X M, Koub S Q, Zhangh Z X, Lindgvistg P A. 2004 The flattened Brazilian disc specimen used for testing elastic modulus, tensile strength and fracture toughness of brittle rocks analytical and numerical results. *Int. J. of Rock Mechanics and Mining Sciences* **41 (2)** 245-253

[5] Golstev V I, Grigoriev, Osintsev A V, Plotnikov A S, Ermakova N S, Gribov N Testing of small-sized samples with the Brazilian test for assessment of brittle strength of metallic materials *IOP: Conference Series (Current issue)*

IOP Conf. Series: Materials Science and Engineering **218** (2017) 012005 doi:10.1088/1757-899X/218/1/012005

Static-dynamic methods of compression of «Al-B» powder composition «Al-B» in the deformable tubular casing

V A Glushenkov[1], V A Mironov[2], I A Belyaeva[1], A E Burmistrov[1] and I S Usherenko[3]

[1] Samara State Aerospace University, Moskovskoye Ave. 34, Samara, 443086, Russia
[2] Riga Technical University, 1 Kalku Street, Riga LV-1658, Latvia
[3] Belarusian State Technological University, Sverdlova str. 13a, 220006, Minsk, Belarus

Abstract. A new scheme of hybrid technology of compacting powders is proposed, which simultaneously use two sources of loading: static and dynamic. Compaction of «Al-B» powder compound is carried out in a metal casing by its longtitudal drawing and radial magnetic pulse crimping. The results of the search experiment are presented, confirming feasibility and efficiency of proposed technical solution. Developing technology of compacting powders will allow producing long-length products.

1. Problem statement

Technical progress of modern engineering requires development and use of new materials for construction of sensitive products for special purposes. An example of such material is the aluminum-boron composite. After the use of the products made of such materials there is a question of the disposal.

One solution to this problem is to crush fragments of products, grind them further into powder with a particle size of up to 100 nm, press them and sinter to obtain products of new application. E.g., the use of the powder composition «Al-B» is suggested for producing grinding tools, elements of radiation protection and other purposes.

The main technological process in the described processing chain of manufacturing new products is the process of powder pressing. Various methods [1÷4] can be used for pressing powders, including use of pulsed magnetic field (PMF) during deformation of the tubular casing filled with powder. However, existing compression methods are not perfect and have certain disadvantages relating primarily to the quality of densification, technology and equipment. That is why the task of developing new compression methods remains relevant.

2. Proposed technical solution

Hybrid technology is proposed that combines static and pulse magnetic loading. Figure 1 shows developed technological schemes of implementation of this technology. The essence of the proposed method is to combine the operations of drawing thin-walled tube filled with powder (static loading) and radial magnetic pulse compression of it (dynamic loading). The following variants of sequence of these operations are possible:

a) magnetic pulse crimping followed by drawing;
b) drawing followed by magnetic pulse crimping;
c) simultaneous drawing and magnetic pulse crimping

IOP Conf. Series: Materials Science and Engineering **218** (2017) 012005 doi:10.1088/1757-899X/218/1/012005

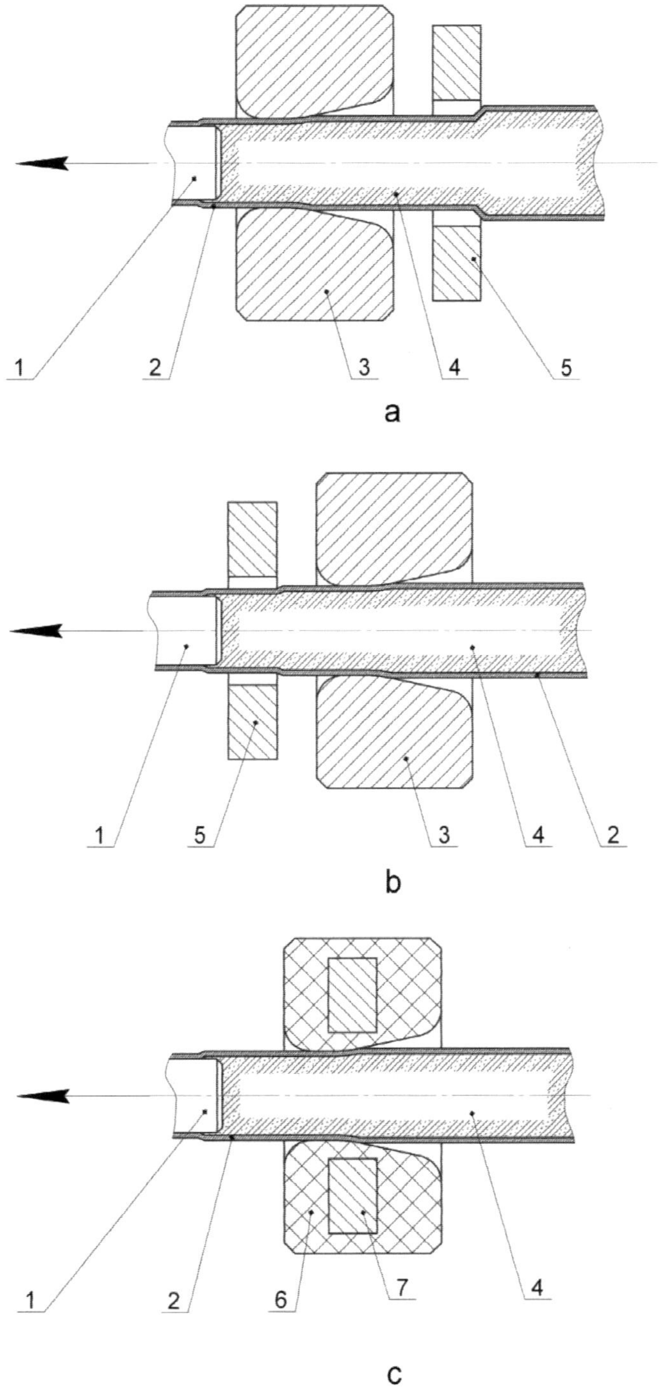

Figure 1. Hybrid schemes of powder pressing, combining static and pulsed magnetic loading: (a) magnetic pulse crimping performed before drawing; (b) magnetic pulse crimping performed after drawing; (c) magnetic pulse crimping and drawing performed simultaneously.

The implementation of these schemes is determined by the placement of the inductor (5) - in front of, behind or inserted directly into the draw die. Inductors are connected with the magnetic pulse in-

IOP Conf. Series: Materials Science and Engineering **218** (2017) 012005 doi:10.1088/1757-899X/218/1/012005

stallation (6). Varying the location of the inductor allows changing the deformation zone, the stress-strain state of the tube and the powder in and scheme of densification.

The on-off ratio of magnetic pulse impact must be linked to a drawing speed so that the deformation zones overlap. This is particularly important for technological schemes a) and b).

3. Search experiment method

A search experiment was conducted to test the feasibility of the proposed technical solutions and evaluate the levels of factors that determine the process.

In the experiment, a powder composition was used with the percentage composition presented in Table 1. For the For the tubular casing a copper (MO) ∅ 18 mm tube was selected with the wall thickness of 1.9 mm and the working length $l_0 = 50$ mm. On one end of the tube, a plug was fixed, used at the same time as swaged before the drawing end of the tube.

Table 1. Chemical elements and their weight percent in the used powder composition.

Spectrum of chemical elements	B	O	M_g	Al	T_i	Mn	Fe	Co	W
Mean value (%)	83.35	5.49	0.53	7.12	0.04	0.08	0.45	0.21	2.71
Standard deviation	0.40	0.04	0.01	0.09	0.02	0.01	0.05	0.03	0.28

The tube was filled with powder and plugged on the other end. Assembling the tube with the plugs was carried out using magnetic pulse crimping on the facility at the magnetic-pulse installation MIU-10VCh with energy discharge value W = 2.6 kJ. The appearance of the sample is shown in Figure 2.

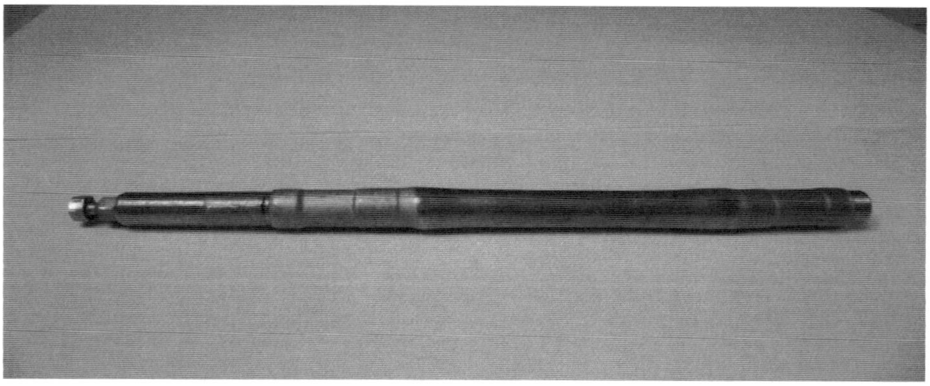

Figure 2. A sample prepared for the experiment.

Steel draw dies used in the experiment had the diameter ∅ 16 mm and 17 mm. The drawing process was carried out on the hydraulic testing machine TsDMU-30. Magnetic pulse crimping was carried out on the magnetic pulse installation MIU-50 at the following main parameters: $C_i = 253,3$ μF; $L_i = 110$ mH; $f_0 = 30$ kHz.

Sintering of the sample was conducted at the temperature $T = 600^0$C and thermal exposure of 60 min. after the furnace reached the specified mode.

Metallography studies were carried out with an electron microscope TESCAN, VEGA series, modification SB.

IOP Conf. Series: Materials Science and Engineering **218** (2017) 012005 doi:10.1088/1757-899X/218/1/012005

4. Search experiment results

The dimensions of the samples after drawing and magnetic pulse crimping are presented in table 2.

Table 2. The dimensions of the samples after drawing and magnetic pulse crimping.

№	Before the drawing		After the drawing		The volume of the powder in the control portion, % to the original volume	After the magnetic pulse crimping	
	D_0 (mm)	L_0 (mm)	D_1 (mm)	L_1 (mm)		W (6 kJ)	$D_{fin.}$ (mm)
Draw die Ø 17mm							
1				52.4		3.8	16
2	17.9	50	17	52.3	93.1	5.3	15.7
3				52.3		7.1	15.4
Draw die Ø 16mm							
4				57.7		3.8	15.7
5	17.9	50	16.2	57.7	91	5.3	15.4
6				57.7		7.1	15.0

The volume of the powder was controlled during the process of pressing. The metallography analysis results are shown in Figures 3 and 4.

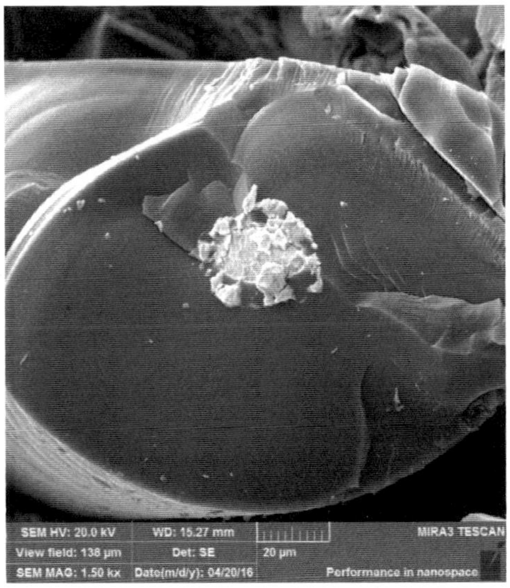

Figure 3. Structure of boron fiber. Tungsten is in the center.

IOP Conf. Series: Materials Science and Engineering **218** (2017) 012005 doi:10.1088/1757-899X/218/1/012005

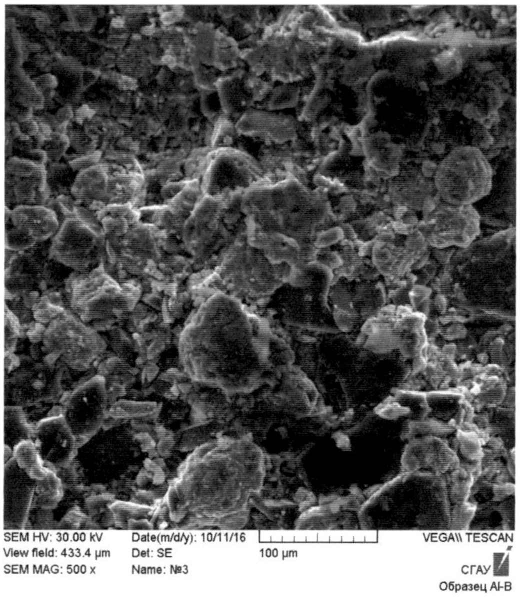

Figure 4. Structure of the powder after the implementation of hybrid technology at different magnification ratio. Dark particles – B, light particles - Al

5. Analysis of results

In the drawing process the powder composition is milling to the size of 50÷20 nm. Moreover, the density of the powder is not uniform across the section. Higher density was observed at the external surface, i.e., at the side of powder exposure to the metal casing.

It was established that both loadings contribute to the densification of the powder. Presumably, the mechanism of densification differs for longitudinal and radial deformation. The density of the powder after the hybrid treatment amounts to 91÷93%.

6. Conclusion

1. A new method of compacting powder compositions in a metal casing – the hybrid technology combining static and pulsed magnetic loading allows simultaneously performing simultaneously two operations: longitudinal drawing and radial magnetic pulse crimping.
2. Several schemes of implementation of this technologies are proposed, one of which has been tested in the laboratory and has confirmed its feasibility and efficiency.
3. The longitudinal-radial deformation of the powder changes the compression mechanism, providing high density of powder composition up to 91 - 93%
4. The proposed technologies offer a possibility of producing long-length products, as one of the areas of utilization of «Al – B» composite materials.

References

[1] Mironov V 1976 Pulververdichten mit Magnetimpulsen, *Planseebericht fur Pulvermet* **24** 175-190
[2] Mironov V, Wolf H 1980 PulververdichtendurchelektromagnetischeKräfteherstellunsmöglichkeitfürWerkzeugagaktivelements *Fertigungetechnik u.Betrieb* **9** 558-560
[3] Boyko, Mironov V and Lapkovsky V 2007 Property and structure changes of sintered powder materials by the treatment in proceedings of euro pm 2007 **3** 439-444
[4] Soda R, Takaqi K, Ozaki K 2016 Numerical simulation of magnetic-aligned compaction with pulsed high magnetic field. *Scripta Materialia* **120** 41-44
[5] Glushenkov V A 2013 Inductors for magnetic pulse processing of materials (Samara) 148
[6] Jusupov P E, Glushenkov V A 2013 Installations for magnetic pulse processing of materials (Samara) 213

IOP Conf. Series: Materials Science and Engineering **218** (2017) 012006 doi:10.1088/1757-899X/218/1/012006

Testing of small-sized disk samples obtained by spark-plasma sintering and electric pulse exposure

V I Goltsev, A V Osintsev and A S Plotnikov

National Research Nuclear University MEPhI, Kashirskoe shosse, 31, Moscow, 115409, Russia

Abstract. The article presents results of design analysis of stress and strain in the disk sample loaded by the Brazilian test (disk compression in the center plane). The research established correlations of the diameter and thickness of the sample acceptable to obtain reliable values of material strength in response to stress. Presented a strong relation of average tensile stresses in the central area of the disc to the stress parameter used to determine the strength of brittle rock materials

1. Introduction

Technology adjustments for production of advanced materials, sintering and welding sintering conditions [1, 2] are usually carried out on the small samples, which diameter size often doesn't exceed 10 mm to 15 mm, and thickness may vary from 1 mm to 10 mm. Standard mechanical test methods are often not applicable for such samples. To evaluate the mechanical properties of the material it becomes necessary to resort to alternative test methods. Examples of such methods are bending of thin discs on a ring bearing [3] and the disk compression in the center plane by the Brazilian test scheme [4, 5]. It should be noted that these methods are used for testing of brittle materials. In particular, the Brazilian test is used to determine the tensile strength of rock. The aim of this study is to investigate the possibility of applying the Brazilian test of small-size samples for evaluating the tensile strength of metal materials with low ductility.

According to ASTM D3967-95a, the tensile strength of the brittle rock materials can be established during center plane compression test on the disk sample with a diameter no less than 50 mm and thickness to diameter ratio in the range of 0.75 to 0.2.

In order to investigate the possibility of using this test method for determining the tensile strength of metal materials with low ductility a numerical modeling of disk compression process was conducted using finite element method (FEM).

2. Parameters of the design model

The simulation of the disk sample loading process was carried out in a verified [6, 7] simulation software complex ANSYS Mechanical 16.2. The design model is a basic symmetry element of compressed disk – a 1/8 part (Figure 1). The thickness of the disc was ranged from 1 mm to 10 mm at the same diameter of 10 mm. The design model contains approximately 332 thousand nodes in the 80 thousand hexahedral elements of the second order (with quadratic shape functions) for a 10x5 mm disc. Material of the model is assumed to be perfectly elastic. The elastic modulus of the material is assumed to be 100 GPa.

IOP Conf. Series: Materials Science and Engineering **218** (2017) 012006 doi:10.1088/1757-899X/218/1/012006

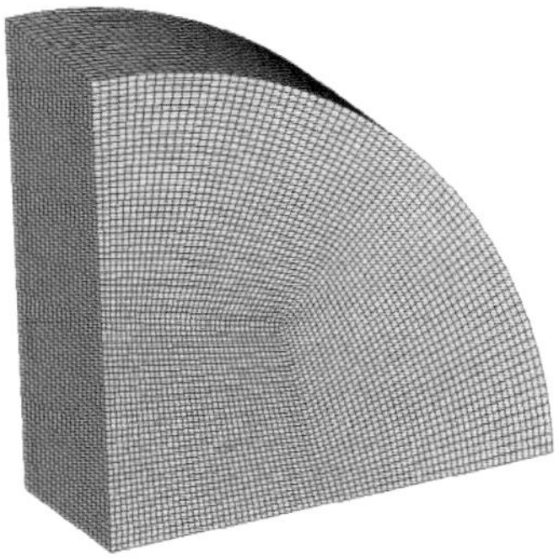

Figure 1. The finite elements grid of the computational domain.

In the model there are boundary conditions imposed of the symmetry on the clipping planes of the computational domain out of the full model. Loading of the model was made by compression of an absolutely rigid plane in the diametrical direction. The elements of the contact interaction were generated for the loading on the cylindrical surface of the disk and on the load plane. Hexahedral elements of second order SOLID186 were used in the formation of the model, as well as elements of the contact interface and TARGET170 CONTACT174 (terminology of [6]). The augmented Lagrange multiplier method was used as the contact algorithm.

The loading was conducted by assigning equal prescribed surface displacements to the surface nodes of the rigid punch. Taking into account the symmetry, total displacement corresponds to two times current displacement and effort corresponds to four times total force of the punch removed from the nodes. The final displacement of the punch amounted to 50 mm, which corresponds to the 0.1 convergence of punches.

In a nonlinear setting, the decision was made for a series of steps - increasing values of prescribed displacements with searching for the equilibrium state at each step by the Newton-Raphson algorithm and handling rigidity matrix by the preconditioned conjugated gradients method.

3. Results the analysis of the FEM modeling

Figure 2 shows the distribution pattern of the first and the third principal stresses and maximum shear stresses corresponding to the maximum model introduction of the punch - 50 microns (0.1 mm closer). The stress values are presented only for the purpose of illustrating their distribution.

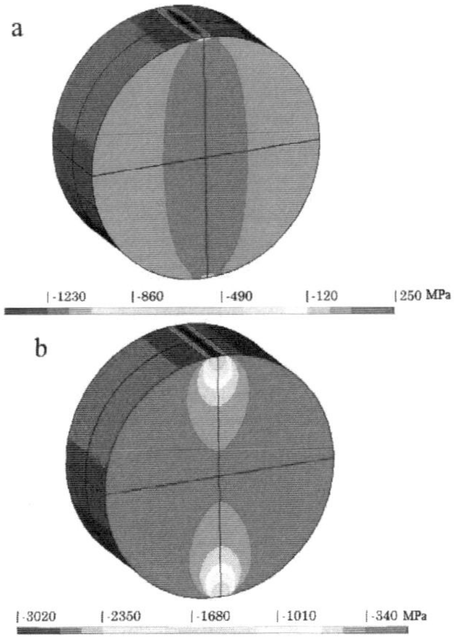

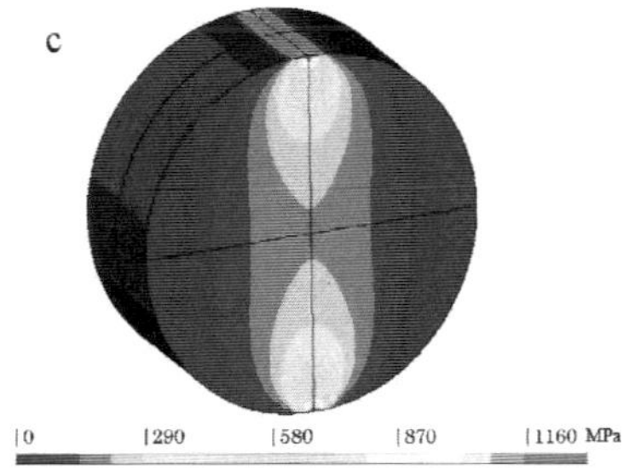

Figure. 2. Distribution of stress state component: (a) – σ_1, (b) – σ_3, (c) – τ_{max}

Presented distributions are characteristic for the solution of described problem almost in the entire range of loading, with the exception of the small area of the initial load. Mostly compressive stresses are active on the main areas of the whole volume of the sample. The exception is the ZY plane of symmetry. On this and nearby parallel planes first principal stresses are positive and has the same direction X ($\sigma_1 \approx \sigma_x$) or close to it. It should be noted that maximum compressive stresses are about 1.5 times greater in absolute value than maximum tensile stresses in the plane of symmetry on the disc the surface, and maximum shear stresses at the same points exceed maximum normal stresses. Phenomenologically, when disk samples are tested on center plane compression, fracture occurs along the diameter coinciding with the direction of compression. A comparison of stress distribution pattern with an actual picture of failure leads to the expected conclusion that failure takes place under the action of maximum tensile stresses, although emergence of failure is possible under the influence of maximum shear stresses.

The maximum tensile stresses are localized near the outer surface of the disc and oriented perpendicular to the direction of compression. At the same time location of the maximum point is not constant during the loading process and shifts in the symmetry plane ZY from the initial position, in the vicinity of the contact zone towards the center of the disc. Localization of maximum tensile stress near the free surface appears to be related to the edge effect. The maximum point doesn't have constant location. The border layer of the disk works in the conditions of flat tension caused by the edge conditions. Accordingly, the stress distribution pattern may depend on the diameter to thickness ratio of the disk during tests. At the same time the ASTM D'3967-95a standard prescribes defining failure stresses by the formula:

$$\sigma_t = \frac{2P}{\pi LD} \tag{1}$$

where P – loading that caused failure, L – sample thickness, D – sample diameter. The stress parameter introduced by the standard should be compared with first principal stresses acting in the whole volume of the sample, in the geometric center of the sample volume and in the geometric center of the circular surface of the sample. Figure 3 shows graphs of correlation between these stresses and the stress parameter σ_t. For clarity of the figure the results are provided for only 4 samples of different sizes, the characteristics of elasticity of the material are assumed to be the same. The horizontal axis represents the stress parameter σ_t , calculated using the formula (1) on the basis of force acting upon the sample, and the vertical axis represents the estimated value of first principal stresses.

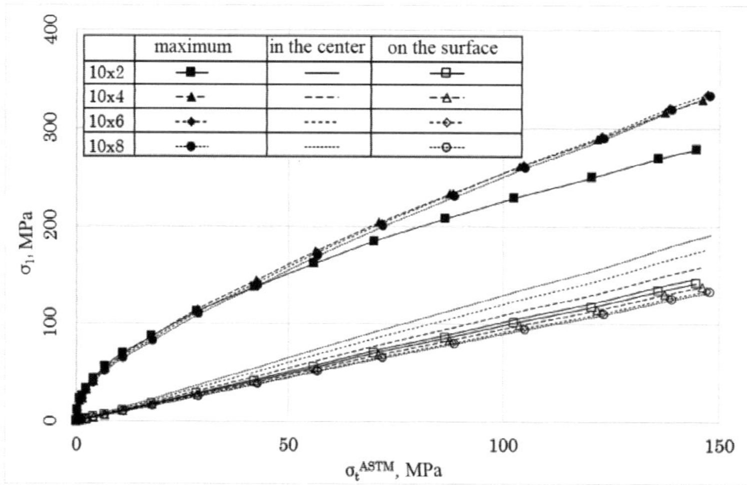

Figure 3. Correlation between the stress parameter σ_1 in the different points and σ_t with varying the thickness of the disk.

These results suggest the following conclusions: σ1max has nonlinear dependence on σt, it is nearly a parabolic branch relative to the σt axis and it depends on the disk sizes ratio – disks size of 10x1 mm, of 10x2 mm and 10x3 mm (the figure shows the curve for the 10x2 disk) have distribution different from all other sizes: dependences of σ1 on σt in the center of the volume and in the center of the surface of the sample are close to the linear and in strong relation with the σt parameter over the entire range.

Consequently, if the failure of the material occurs when maximum tensile stresses reach a certain limit, then the amount of the stress limit does not match the value of the parameter σt assessed by the ASTM D3967-95a method. If the failure occurs on reaching average stresses in a section of limit value, which is characteristic of heterogeneous materials such as concrete or rocks then the σ_t parameter is a good estimate of the first principal stress, averaged along the X axis of the disk.

The results of computational analysis were compared with the results of the test of gray cast iron disks. [8] A strong correlation was discovered between the pressure in the center of the sample and tensile strength of the material.

4. Conclusions

Using the FEM a design analysis was conducted on the stress distribution and deformation in small-sized metal disk samples loaded by the Brazilian test method. Strong correlation was presented between the normal stress acting in the center of the disc and the design parameter recommended by ASTM standards for determining the tensile strength of brittle materials.

5. References.

[1] Grigoryev E *et al* 2011 *Int. J. Machines Technologies Materials* **8** 8

[2] Bashlykov S S Demenuk.V D Grigoriev E G Olevsky E A Iurlova M S 2013 Electropulse method for consolidation the uranium mononitride powder *Physics and Chemistry of Material Processing* **5** 77-83

[3] Zholnin A G, Kovaleva I V, Pahilo-Daryal I O, Litvinova I S, Grigoryev E G, Olevsky E A, Goltsev V I, Rytenko V R Effect of the size of alumina powder particles on the results of spark-plasma sintering *Physics and Chemistry of Material Processing* **2** 5-10

[4] Wang Q Z *et al* 2004 *Int. J. of Rock Mechanics and Mining Sciences.* **41** **(2)** 245

[5] ASTM D3967-95a. Standard Test Method for Splitting Tensile Strength of Intact Rock Core Specimens

[6] Release 16.2 Documentation for ANSYS [electronic document] ANSYS Inc. Electronic data and software (104019 files: 10660130531 bytes)

[7] NAFEMS search engineering analysis and simulation *FEA, Finite Element Analysis, CFD, Computational Fluid Dynamics, and Simulation* (NAFEMS Ltd., Hamilton, United Kingdom) Date of access: 28.01.2016. URL: http://www.nafems.org/search/ansys/0/9///

[8] Golstev V I, Grigoriev, Osintsev A V, Plotnikov A S, Ermakova N S, Gribov N Testing of small-sized samples with the Brazilian test for assessment of brittle strength of metallic materials *IOP: Conference Series (Current issue)*

IOP Conf. Series: Materials Science and Engineering **218** (2017) 012007 doi:10.1088/1757-899X/218/1/012007

Testing of small-sized samples with the Brazilian test for assessment of brittle strength of metallic materials

V I Golstev, E G Grigoriev, A V Osintsev, N S Ermakova, A S Plotnikov and N A Gribov

National Research Nuclear University MEPhI, Kashirskoe shosse, 31, Moscow, 115409,Russia

E-mail: natalia_ermakova2503@mail.ru

Abstract. The article provides an experimental basis for application of the disk sample loaded by the Brazilian test method (disk compression in the diameter plane) for the brittle strength assessment of the metallic materials during testing of the small-sized samples. The research presents a strong correlation between strength of the cast iron, assessed by the testing of disk samples and the tensile strength. The method was used to study the mechanical properties of heavy tungsten pseudo-allows obtained by exposure to high-voltage electric pulse.

1. Introduction

Testing of modes of electropulse impact in production of advanced materials is generally carried out on small samples, often not exceeding the size of 10 mm 15 mm in diameter, with thickness than may vary from 1 to 10 mm. Standard methods of mechanical testing are not applicable for such samples. This work provides substantiation of the use of Brazilian test on small-size samples to evaluate the tensile strength of metallic materials. An experimental analysis of the test results on cast iron samples was conducted by the method of digital image correlation (DCI).

2. The results of the experimental research

An experimental research was conducted on disk samples of gray cast iron GCI 10-40 ($\sigma_в^р$=104, $\sigma_в^{сж}$=404 MPa). The disks had three types of size *DxL*: 10x4 mm, 10x5 mm and 15x4 mm. To evaluate the mechanical properties of the material the alternative testing methods are have to be used. One of such method can be disk compression in the center plane by the Brazilian test [1, 2], which is used for the indirect determination of tensile strength of rock.

The loading of the samples was carried out on a test machine INSTRON in a special device that provides strict alignment of the planes of the punches. Simultaneously with the loading of the samples, the displacement of the surface points of each samples were registering by the DCI method [3, 4]

The distribution of the strain ε_x in surface points of cast iron sample in the initial stage of the loading and the stage preceding failure is shown on Fig. 1 a, b.

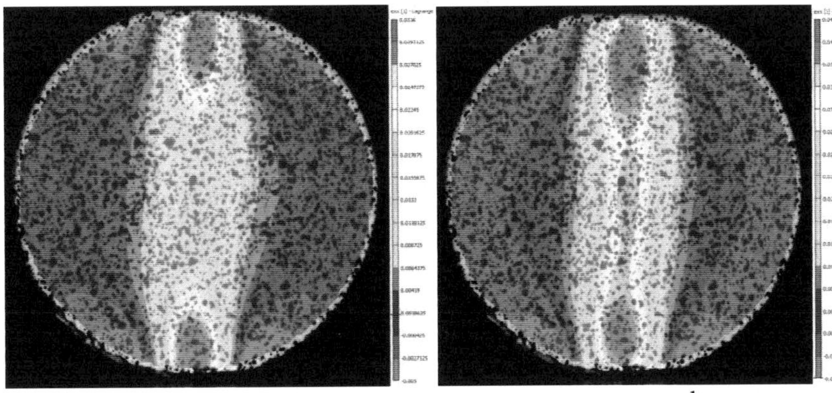

a b

Figure 1. Color images of strain distribution ε_x over the surface points of cast iron: (a) the initial stage of the loading; (b) the stage preceding the failure

Tensile strain distribution ε_x (red zones represent the maximum strain) correspond exactly to the distribution pattern of tensile stresses in the design analysis presented in [5], which indicates the identity of the results obtained by the calculation and experimental methods. The symmetrical strain distribution patterns in respect to the axes X and Y are observed at all the stages of loading the samples. In the surface points plane stress condition presents with the principal tensile stresses σ_x stresses and compressive stresses σ_y, and consequently, significant shear stresses equal to the half the sum of the absolute values of the principal tensile and compressive stresses. The shear stresses cause the appearance of displacement cracks which combined lead to the failure of the sample in the center plane, where maximum tensile stresses apply.

The appearance of such cracks is characteristic of samples with the size of 15x4 mm and not observed for samples with the size 10x5 mm in which the failure starts in the center area of the sample. Stepped fracture in the center plane of the sample (Fig. 2) indicates that the sample failure occurs under the influence of shear stresses. There is no explosive failure typical for brittle fracture caused by normal stresses. A slow withdrawal of one half of the sample from the other is observed, which is usually inherent to plastic failure. The sample deformation diagrams are characterized by the development of plastic flow and by the transition of stress through the maximum.

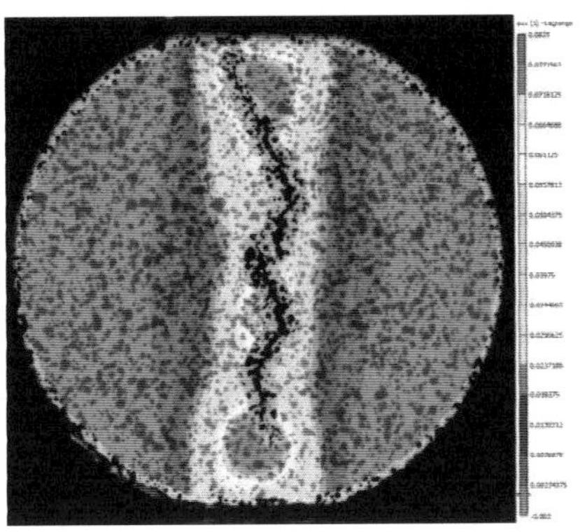

Figure 2. The nature of division of the cast iron sample during its destruction.

Table 1 shows the test results for 9 discs of cast iron: the maximum load corresponding to the start of the failure for all tested samples and damaging stresses calculated by the formulas (1) and (2),

$$\sigma_t = \frac{2P}{\pi LD} \quad (1)$$

$$\sigma = \frac{P}{LD} \quad (2)$$

Table 1. Fracture strength of cast iron, defined by the formulas (1) and (2).

№ sample	Sample size (mm)	P (N)	σ_t (MPa)	σ (MPa)
1	10x4	5333	85,3	134
2	10x4	6028	96,2	151
3	10x4	6979	111	174
4	10x4	6847	109	171
5	10x4	7029	112	176
6	10x5	9009	115	180
7	10x5	8783	112	176
8	15x4	8290	87,9	138
9	15x4	7550	80,2	126
	The average values		101	158,4

The average values of pressure calculated by the formula (1) were close to the values of the tensile strength of the material. Therefore in order to calculate the failure stresses in the testing of metallic brittle disks by the Brazilian test method it is possible to use the formula (1), recommended by the ASTM D3967–95a standard.

3. Testing of disk samples of tungsten heavy alloy VNZh-90 (90W-7Ni-3Fe) obtained by the method of high-voltage electric discharge consolidation

The examined above method of testing disk samples was used to evaluate the brittle strength of the tungsten heavy alloy VNZh-90 (90W-7Ni-3Fe), obtained under various sintering conditions. Table 2 shows the test results for series of samples. Legend to the table: U – pulse voltage; q – pressure during the powder sintering; D and L – diameter and thickness of the sample; P_{max} – the maximum load at which the sample failure occurred or loading was stopped, due to a significant deformation of the sample without its failure; σ_t - stress calculated by the formula (1); ω- displacement of the baseplate of the testing machine considering its ductility. It should be noted that a part of the samples wasn't brought to failure due to the emergence of large plastic deformation or due to formation of delamination and spalling.

Table 2. The testing results for a series of disk samples of tungsten heavy alloy VNZh-90 (90W-7Ni-3Fe)

Sample	U (kV)	q (atm)	D (mm)	L (mm)	$P_{max,}$ (N)	σ_t (MPa)	ω at P_{max} (mm)	Notes
A1-1	5	2	10,8	5,2	17924	203	0,76222	brittle fracture of the sample
A1-2	5,5	2	10,6	4,5	30523	407	1,16487	plastic deformation without the failure of the sample
A1-3	5,5	1,5	10,6	4,5	50881	679	2,88742	brittle fracture of the sample
A1-4	5,6	1,5	10,5	4,5	29863	402	1,05569	brittle fracture of the sample
A1-5								Not tested
A1-6	5,5	1	10,9	5,5	34872	370	1,08068	brittle fracture of the sample
A1-12	4,5	2	10,3	4,0	5991	93	0,58642	brittle fracture of the sample
A1-13	5	2	10,3	5,8	24757	264	1,15809	delamination deformation
A1-14	5,2	2	10,5	5,7	40861	435	1,55914	delamination deformation, spalling
A1-15	5,4	2	11,0	5,6	48736	504	1,98984	delamination deformation, spalling
A1-19	5,6	2	11,0	5,1	34788	395	1,14575	brittle fracture of the sample
A1-20	5,7	2	10,8	5,7	40767	422	1,22498	brittle fracture of the sample
A1-21	5,8	2	11,0	5,0	35741	414	1,24907	brittle fracture of the sample

Analysis of the results shown in Table 2 allows making a clear conclusion about the increasing strength of material at the increasing pulse voltage with pressure amount of 2 atmospheres, as well as about the maximum material strength at considerable plastic deformation under pressure of 1.5 atm and pulse voltage of 5.5 kV.

Figure 3 shows the deformation diagrams for the first five samples tested (see Table. 2). The diagrams also indicate the presence of the plastic flow of the sample prior to its destruction.

Another useful feature of disk samples loaded by the Brazilian test method was discovered during the testing of steel samples. It was found that symmetrical with respect to the axes X and Y strain distribution on the surface of the sample is characteristic. Violation of this symmetry signals a breach in the internal structure of the material which consequently results in premature failure of the sample in the form of spalling of its individual parts. [6]

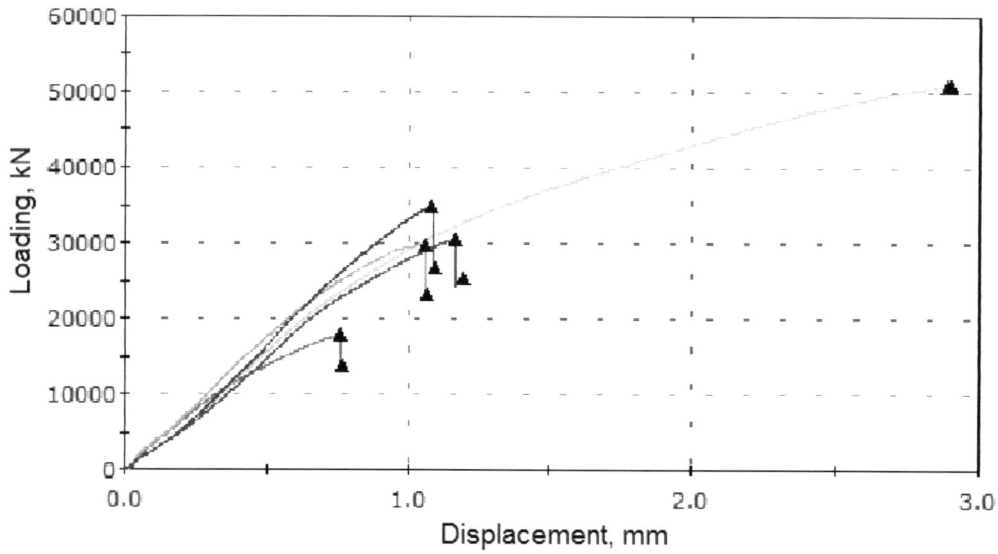

Figure 3. Deformation diagrams for the first five samples of Table 2.

4. Conclusion

This presents the possibility of indirect determination of the tensile strength of metallic materials based on the results of the testing of small-sizes disk samples with the Brazilian test. The test results for the disk samples of the tungsten heavy alloys show high strength and deformation properties of the material under optimal production conditions.

References

[1] *Wang QZ, Jiaa X M, Koub SQ, Zhangh Z X, Lindgvistg P A* 2004 The flattened Brazilian disc specimen used for testing elastic modulus, tensile strength and fracture toughness of brittle rocks analytical and numerical results. Int. J. of Rock Mechanics and Mining Sciences. 41 (2) 245-253

[2] ASTM D3967-95a. Standard Test Method for Splitting Tensile Strength of Intact Rock Core Specimens.

[3] Sutton M A, Orteu J, Schreier H W 2009 Image correlation for shape, motion and deformation measurements: basic concepts, theory and applications. - Springer 321

[4] *Chu T C, Ranson W F, Sutton M A and Peters W H* 1985 Applications of digital image correlation techniques to experimental mechanics *Experimental mechanics* 232-245

[5] Golstev V I, Osintsev A V, Plotnikov A S Testing of small-sized disk samples obtained by spark-plasma sintering and electric pulse exposure *IOP: Conference Series (Current issue)*

[6] Golstev V I, Grigoriev, Osintsev A V, Plotnikov A S 2016 Evaluation of material homogeneity in products obtained by methods of high-voltage electric resistance welding and pulse electric current sintering *Tekhnologiya Mashinostroeniya [Mechanical Engineering Technology]* 7 71-77

Biomechanical properties of composite compact-porous titanium produced by electric discharge dintering

D Minko, K Belyavin and V Sheleg

Belarusian National Technical University, Nezavisimost Ave. 65, Minsk, 220013 the Republic of Belarus

E-mail: dz-m@tut.by

Abstract. The main disadvantage of currently used endosteal implants is their unsatisfactory biostable performance. Under action of functional stress caused by flaws of the design or lower mechanical characteristics the areas of stresses extreme concentration exceeding strength limits of bone tissue appears in the bone surrounding the implant that leads to the tearing away the implant. The problem of specific pressure lowering on the bone and uniform distribution of stress is solved by two ways: the increase of the implant area and the search of implant materials with optimum biomechanical properties. Porous materials of spherical titanium powders have adjustable pore size and large unit surface area, as well as possess high biologic compatibility with living tissue. This allows reduction of the rejection reaction due to a more even stress distribution around the functioning implant. Clinical results show that such implants have more stable physical and chemical properties.

1. Introduction

One of the area of application of composite titanium-based materials in medicine is implantation of artificial devices of various forms and purposes into the human body for replacement, reconstruction or repair of tissues and organs. For instance, according to the World Health Organization an uptake of patients older than 35 on the problems of restoring the function and aesthetics of dental system accounts for 98,8 % of the global population. Most of these complaints are related to the diseases leading to partial or complete loss of dentition, and can be solved by using various prosthetic devices with implants.

A review of scientific and medical literature on endosteal implantation [1-5] shows that the most widely used methods of ensuring human life activity are those based on scientific developments in production of new materials for manufacturing of medical devices.

The main disadvantage of currently used endosteal implants is their unsatisfactory biostable performance. Implants suffer heavy axial and bending stresses that transmit to the supporting bone. Concurrently, imperfect design solutions or wrong choice of materials cause formation of areas in the bone surrounding the implant where excessive stress concentration exceeds the limits of durability of the bone. This results in formation of traumatic capsule of connective tissue around the implant, and consequently, implants rejection. There are two ways of solving the problem of reducing the unit pressure on the bone surrounding the implant and ensuring uniform stress distribution in it: by increasing implant area and by searching for implant materials with optimum biomechanical properties.

The use of devices with a surface layer of porous titanium powder [2] in modern medicine, especially in implant surgery, is conditioned by their important advantages in comparison to currently used cast titanium products with modified surface morphology. The porous layer formed by sintered powder has ramified spatial structure with a reduced modulus of elasticity. By its parameters it is well suited to penetrate bone tissue, allowing fixing the implant in the body of the patient for a lifetime. [3]. Adjustable pore size and large unit surface area allows even distribution of functional stress and reduction of the rejection reaction. Clinical results show that implants with a porous surface have more stable physical and chemical properties [4, 5].

The paper presents the structural and biomechanical properties of the compact-porous material composed of spherical titanium powders obtained by pulsed electric current sintering.

2. Materials and methods of research

One of the possible methods offered for producing the compact-porous material with a porous layer of spherical titanium powders is Electric Discharge Sintering (EDS). The EDS technology is based on a high voltage discharge of a capacitor bank through metal powder, contained in a dielectric mold between the electrode punches [6].

Experimental samples were produced by using commercially pure Russian titanium brand VT1-0 GOST 19807-91, which is close, by its chemical composition, to titanium brand Grades 1-4 ISO 5832/II or ASTM F 67-89 (Table 1).

Table 1. Chemical composition of the titanium brands GOST19807-91 and ISO5832/II (ASTM F 67-89).

Brand\Element	N (%)	C (%)	H (%)	Fe (%)	O (%)	Ti (%)
BT1-0	0,04	0,05	0,008	0,15	0,1	the rest
Grade 1	0,03	0,1	0,0125	0,2	0,18	the rest
Grade 2	0,03	0,1	0,0125	0,3	0,25	the rest
Grade 3	0,05	0,1	0,0125	0,3	0,35	the rest
Grade 4	0,05	0,1	0,0125	0,5	0,5 (0,4)	the rest

Spherical titanium powders produced by plasma dispersion in the vacuum of the rotating consumable electrode [7] had a nearly ideal spherical shape (Figure 1).

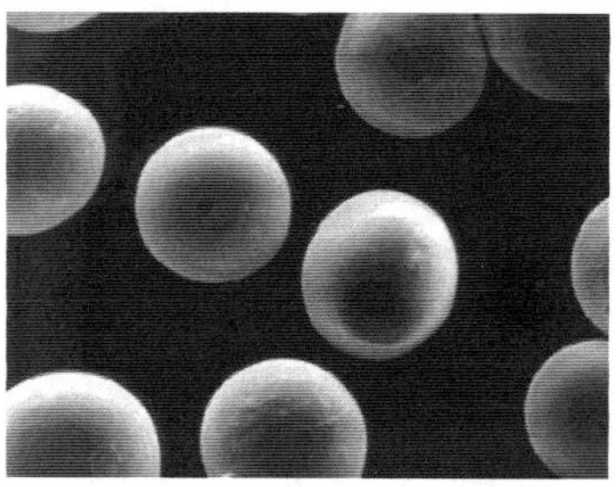

Figure 1. Particles of a spherical titanium powder, x50.

The experiments were conducted using the powders divided into fractions with particle size (0.16-0.2) mm, (0,2-0,315) mm and (0,315-0,4) mm. The powders of the selected fractions were used to produce cylindrical samples 6 mm in diameter and 18 mm length (Figure 2).

A distinct feature of obtaining the implant material samples by the developed technology is the use of EDS as an intermediate operation for preliminary formation of the porous layer. The final processing was conducted by a low-voltage pulse of electric current. This technology allows increasing the strength of the samples, the final sintering process is carried out at lower temperatures without the use of refractory forms and adjuvants. The concurrent shrinkage of samples is minimal, which allows avoiding the cracking and maintaining the accuracy of the geometrical sizes and forms [8].

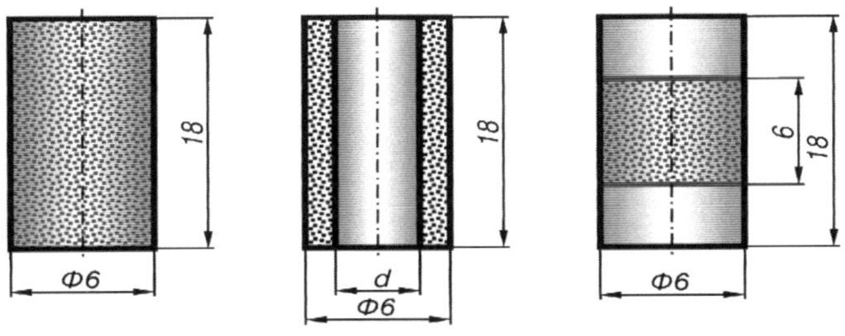

a b c

Figure 2. Sizes of the experimental samples of a porous (a) and compact-porous material with an external (b) and internal (c) porous layer.

The microstructure of the powders and porous samples was studied on a metallographic microscope Polyvar (Reichert, Austria). The shape and the surface morphology of the powders were studied with a scanning electron microscope CamScan (Oxford, UK). The porosity of the experimental samples was determined by hydrostatic weighing. Studies of porosity distribution, pore size and the size of the contacts between powder particles in different sections of the samples were conducted in a software complex for image processing and analysis AUTOSCAN (Spectroscopic System, Belarus). The bending strength testing of the experimental samples was conducted on a universal testing machine model 1195 (Instron, UK) with the use of special tools.

3. Discussion of results

3.1. Structural properties of the porous layer

In experimental samples, there are clearly visible pores exposed over the entire surface of the powder layer formed by spherical titanium particles. (Figure 3, a).

Structural studies have shown that the porous layer has through porosity, uniformly distributed over the cross-section, with no dead-end pores.

Porosity of the obtained samples is in the range of 37% to 40% (Table 2), suggesting the absence of particle deformation and not entirely dense packing (minimum theoretical porosity of spherical powder particles of the same size forming a close-packed structure is 25.9%, maximum in free cubic packing is 47.6%). Consequently, the values of porosity for each fraction are nearly identical. The studies show that the porous layer with thickness of 2 to 3 particle diameters allows increasing by a magnitude of the effective surface area of implant interaction with the surrounding tissue without increasing its geometric dimensions. On the fractography images of the brittle fracture of the samples there are clearly visible good contacts, both between individual titanium particles in the porous layer, as well as between the particles of the porous layer and compact titanium (Figure 3, b).

a b

Figure 3. The external appearance (a) and morphology of a brittle fracture (b) of the experimental samples of a compact-porous titanium.

For production of endosteal implants of various designs and applications, the important characteristic that defines their consumer properties is the size of the pores where the ingrowth of surrounding tissue occurs. Research studies show [3-5] that the pore size ensuring germination and good bone implant fixation should exceed 100 microns. For practical purposes the pore size greater than 100 microns is recommended for extensive bone ingrowth and strong interface fixation to occur. The desired pore size can be achieved by using powders with different particle size (Table 2).

Table 2. Structural properties of a porous layer of the samples.

Particle size (mm)	Porosity (%)	Average pore size (μm)
0,16-0,2	37-38	56-80
0,2-0,315	37-38	80-136
0,315-0,4	38-39	127-198

3.2. Biomechanical properties

Duration of service for the implants in the human body largely depends on the mechanical properties of the materials used for their production. The elastic modulus of compact titanium and its alloys (110-112 GPa) is closest to the elastic modulus of bone (10-30 GPa) [9]. Porous layer of spherical titanium powders has lower density characteristics comparing to ones of compact material but is has nearly the same range of elastic modulus as bone tissue. (Table 3). This distinct feature of the porous titanium presents various opportunities for optimization of biomechanical properties in new constructions of surgical implants.

Table 3. The mechanical properties of a porous layer of the samples.

Particle size (mm)	Elastic modulus (GPa)	Compressive strength (MPa)
0,16-0,2	38-41	194-198
0,2-0,315	27-30	145-149
0,315-0,4	17-21	47-51

Since endosteal implant structures typically use parts made of composition of porous and compact titanium, a research was conducted to evaluate the durability of connection of porous compact layer to the compact part of samples on a flat surface (table 4).

Table 4. The strength of a porous and compact titanium coupling.

Particle size (mm)	Shear strength (MPa)	Tearing strength (MPa)
0,16-0,2	81-85	37-43
0,2-0,315	72-76	26-29
0,315-0,4	49-53	17-19

Using of compact-porous titanium as an implant material allows a very flexible selection of their elastic characteristics in accordance with the existing bone structure. The module of the elasticity of the implant may vary in length depending on the ratio s/S, which defines the relative cross sectional area of the compact part of the implant (Figure 4).

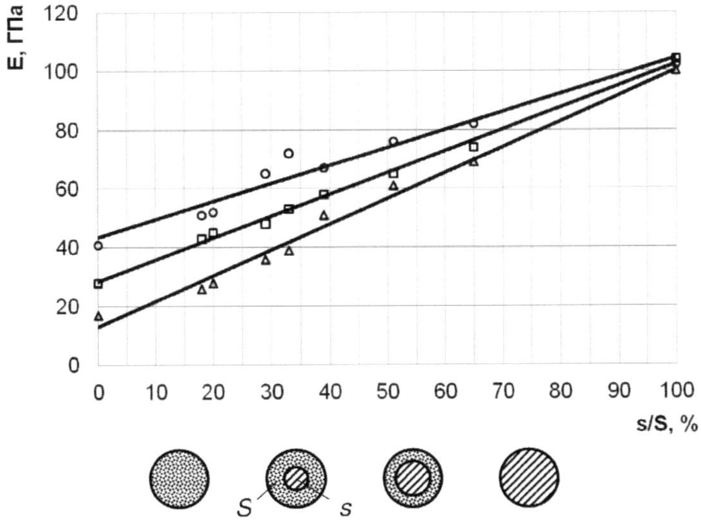

Figure 4. The dependence of a compact-porous titanium modulus of elasticity from the correlation of cross-sections

● – (0,16-0,2) mm;
■ – (0,2-0,315) mm;
▲ – (0,315-0,4) mm

The main parameter for biomechanical assessment of strength of implant bonding with the bone (osseointegration) is shift pressure or amount of force per unit of implant surface area, which leads to the detachment of the implant from the bone. This index depends on the ratio of strength of the implant displacement to the area of its lateral surface. Comparative experimental studies conducted on fragments of animal bones with implanted samples made it possible to define shift pressure for titanium samples with various surface textures, including smooth, corrugated, helical and porous, 6 months after implantation (Table 5). The maximum cracking pressure on the porous layer of powder particles with size of (0,315-0,4) mm was 22 MPa, which exceeds the value of unit cracking pressure of implants with a smooth surface by 3.9 times, with ribbed - by 1.8 times, with threaded – by 1.2 times.

4. Application of composite compact-porous titanium.

Equipment and mechanisms for production of endosteal implants of various shapes and sizes were made based on the developed technology. Technology application allows forming porous layers one or two powder layers thick, on devices the size range of 2 mm to 100 mm.

The conducted technical, biomedical and clinical trials have shown that the compact-porous titanium implants have biomechanically optimal density and structural characteristics, does not have general toxic, irritating and allergenic effects on the body, shows a high degree of biocompatibility and osseointegration (Figure 5).

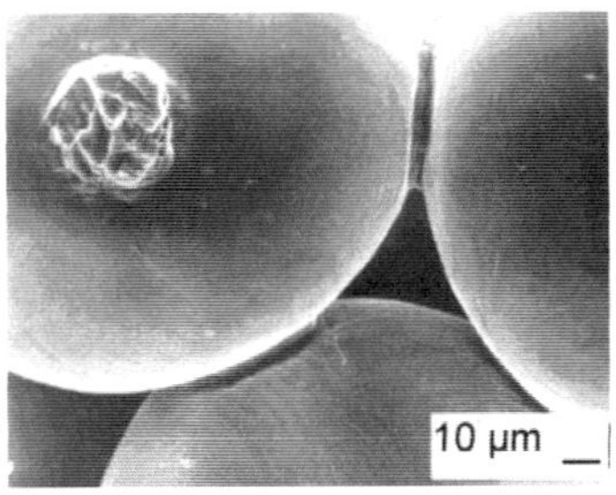

a

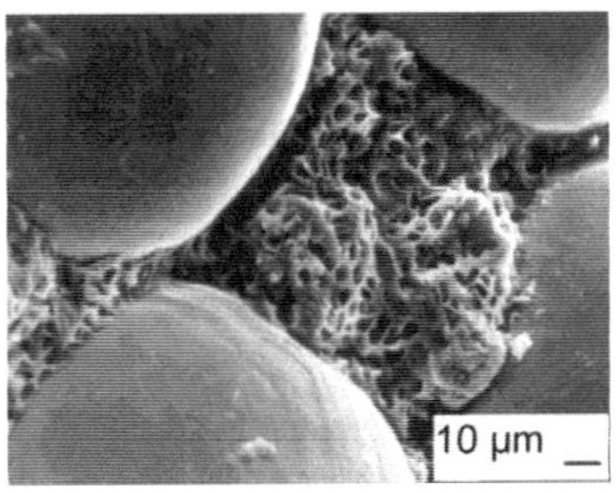

b

Figure 5. The structure of the porous layer of spherical particles of titanium (a) and a bone tissue germinated into it (b).

Designs of endosteal implants made of compact-porous titanium has been developed for use in dentistry, maxillofacial surgery, traumatic surgery (Figure 6), and other fields of medicine.

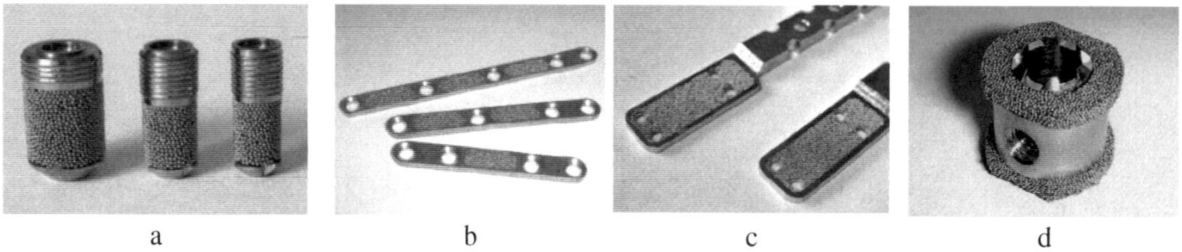

Figure 6. Medical devices made of compact-porous titanium dental implants (a), plates for the fusion of bones (b), implants of low jaw (c), and an implant for interbody fusion of the vertebrae (d).

5. Conclusions

The use of porous composite porous-compact materials in implants shows that the endosteal implants with a porous layer of spherical titanium powder possess high biological stability in the body and can be implanted for life. The results of twenty years of clinical studies show that large unit surface combined with small geometric sizes, high mechanical strength, good adaptation to the bone structure and biocompatibility provide porous-compact implant materials advantages in durability compared to the monolithic materials with surface modified by various technologies.

Clinical results

The above-described endosteal implants with a porous layer of spherical titanium powder successfully passed the certification tests and has been approved for serial production and use in medical practice in the Republic of Belarus.

References

[1] Martin R B and Paul H A 1988 Effects of estrogen deficiency on the growth of tissue into porous titanium implants *J. Of Bone and Joint Surgery* **70-A(4)** 540-547

[2] Story B J and Wagner W R 1998 New enhanced coating for dental implants *Sulzer Technical Review* **1** 38-40

[3] Jian Xu, Xiao-Jun Weng, Xu Wang, Jia-Zhang Huang, Chao Zhang, Hassan Muhammad, Xin Ma and Qian-De Liao 2013 Potential use of porous titanium–niobium alloy in orthopedic implants: preparation and experimental study of its biocompatibility *In Vitro Plos One* **8(11)** 1-13

[4] Pilliar R M, Lee J M and Maniatopoulos C 1986 Observations on the effect of movement on bone ingrowth into porous-surfaced implants *Clin. Orthop. Rel. Res.* **208** 108-113

[5] Pilliar R M 1986 Implant stablization by tissue ingrowth. *Proceedings of an International Congress* ed D van Steenberghe 60-76

[6] Minko D and Belyavin K 2016 A porous materials production with an electric discharge sintering *Int J of Refractory Met and Hard Mat* **59** 67-77

[7] Glazunov S G and Govorov V G 1976 An installation (VGU-2) for the manufacture of spherical powders (granules) from chemically active metals and alloys *Powder Metallurgy and Metal Ceramics* **9** 84-90

[8] Belyavin K E, Minko D V and Reshetnikov N V 2006 A new technology of hardening porous materials of titan powders *Powder Metallurgy World Congress* **2** 1012-1013

[9] Yamanda H 1973 Strength of biological materials 342

Consolidation of metallic hollow spheres by electric sintering

V Mironov, A Tatarinov and V Lapkovsky

Scientific Laboratory of Powder Materials, Riga Technical University, Kipsalas Str. 6B-331, LV-1048, Riga, Latvia.

Abstract. This paper considers peculiarities of the technology of production of structures from metallic hollow spheres (MHS) using magnetic fields and electric sintering. In these studies, the raw material was MHS obtained by burning of polystyrene balls coated by carbon steel. MHS had an outer diameter of 3-5 mm and a steel wall thickness of 70-120 microns. Pulsed current generators were used for electric sintering of MHS to obtain different spatial structures. Since MHS have small strength, the compressive pressure during sintering should be minimal. To improve the adhesion strength and reduce the required energy for sintering, hollow spheres were coated with copper by ion-plasma sputtering in vacuum. The coating thickness was 10-15 microns. The ferromagnetic properties of MHS allowed using of magnet fields for orientation of the spheres in the structures, as well as using of perforated tapes acting as orienting magnetic cores. Ultrasonic testing of MHS structures has been tried using through propagation of ultrasound in low kilohertz frequency range. Sensitivity of the propagation parameters to water filling of inter-spheres space and sintering temperature was demonstrated.

1. Introduction

Structures from metal hollow spheres (MHS) are among many new promising cellular structures designed for various applications [1]. Such structures are used as filters, catalysts, and exhaust noise absorbers. Their advantages are low weight, high sorption capacity and other special properties. The raw materials for such structures are MHS typically made of steel. For the production of these structures, different bonding techniques were implemented including brazing, sintering, and infiltration of [2-4]. Technology of consolidation of MHS by electric heating with a welding machine was considered [3]. The technology of consolidation of powder materials introduced in the Riga Technical University applies electrical generators of pulsed discharge currents [5-7]. The complications of using this technology for MHS consolidation were: the need to apply the technological shell (coating) made of a material with a high electrical conductivity, non-uniformity of the material distribution during compaction, limited shapes and sizes of products.

Besides, free unsupported filling of the material doesn't provide the desired product quality. Thus, the search for new ways to improve quality by perfection of all operations of the electrical sintering process is needful. The need for non-destructive quality control of the sintered bodies is also an urgent task which may have a non-trivial solution because of the specific properties of MHS structures, in particular its extremely high porosity. The ability to use ultrasonic attenuation to characterize structural density in metal foams was showcased by the laser ultrasonic detection method [8]The presence of adhesion and the quality of consolidation was tested by ultrasound in a variety of practical applications [9]. However, these results are not expandable on MHS structures.

2. Experimental procedure

Experimental studies were conducted on MHS made of steel by the technology developed in Fraunhofer IFAM Department of Powder Metallurgy and Composite Materials, Dresden [4]. MHS were obtained by burning of polystyrene balls coated by carbon steel. It had an outer diameter of 3-5 mm and a steel wall thickness of 70-120 microns. The general view of MHS and the structure of its surface are shown in Figure 1 and Figure 2.

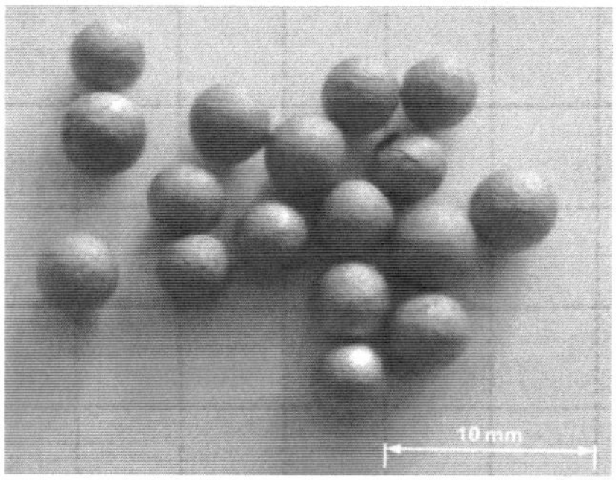

Figure 1. Steel MHS diameter of 3-5 mm obtained from Fraunhofer IFAM

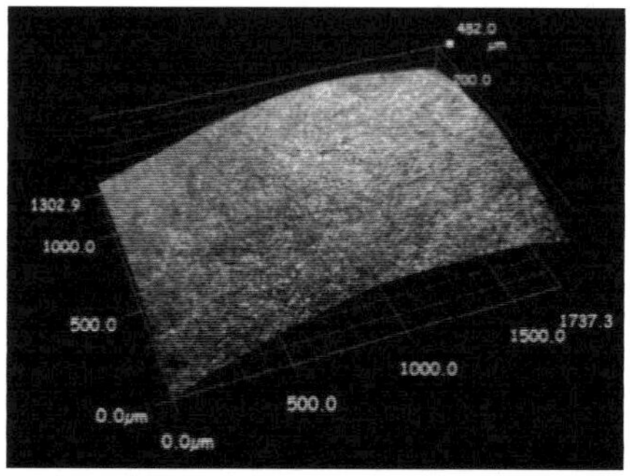

Figure 2. 3-D digital optical microscopy of MHS surface.

Electrical sintering was performed by a pulsed current generator with the operating voltage 200-400 V and power consumption of storage capacitor 1-2 kJ. At the discharge moment, a pulsed discharge current of 0.5 - 1, 2 kA with duration of 10-50 ms was passed through a volume filled by layers of MHS. The axial compression of the material was provided by a flat inductor exited by the pulsed generator. The applied force reached 150 N. The electrical sintering process combined with electromagnetic compaction is schematically shown in Figure 3.

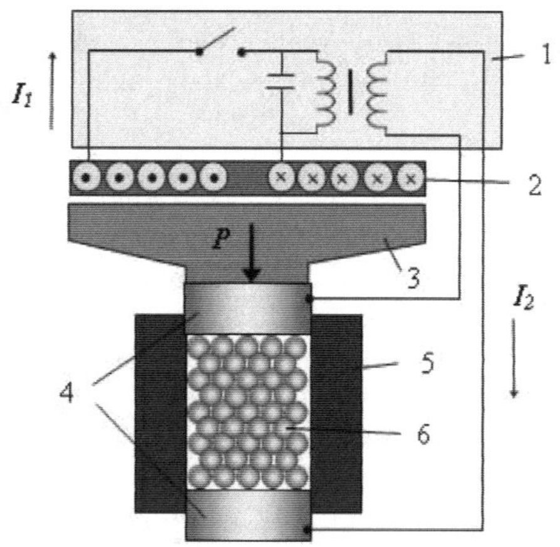

Figure 3. Schematic layout of combined method of electromagnetic compaction and electrical sintering:: 1 – pulsed current generator ; 2 – inductor; 3 – punch; 4 – electrical poles; 5 – ceramic mold; 6 – sintered sample of MHS.

3. Results and Discussion

To obtain a material with stable projected properties different approaches were tested. One option was modification of MHS surface by ion-plasma sputtering and providing a 15-30 microns layer of copper coating (Figure 4-Figure 6). Applying of a thin layer of copper on the surface of steel spheres allowed better sintering of MHS due to increased electrical conductivity of the contact layers and reduction of the discharge current. Good sintering was obtained at a current value within 300A [10].

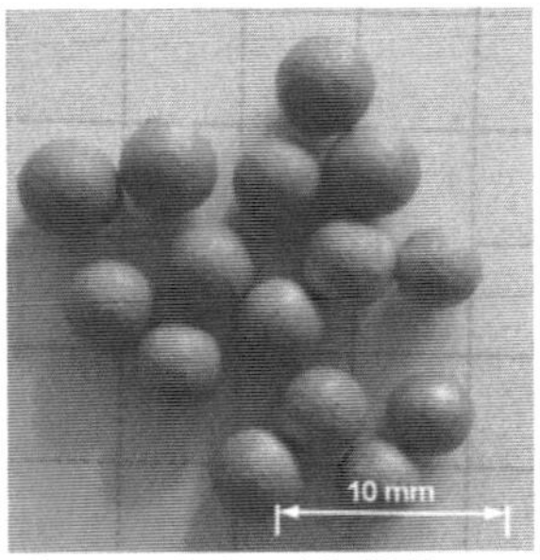

Figure 4. Steel MHS coated by copper.

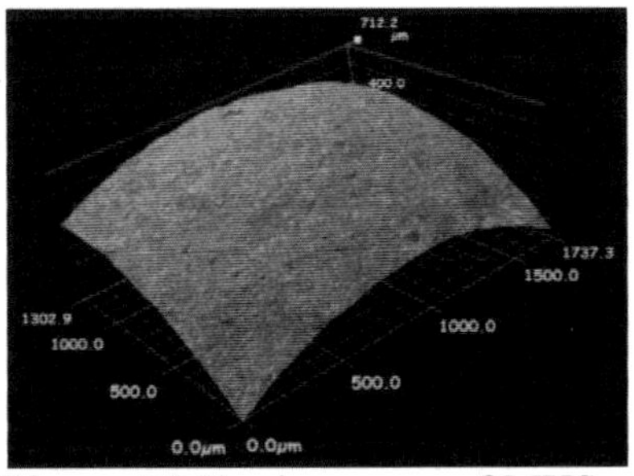

Figure 5.3-D microscopy of MHS surface after copper coating.

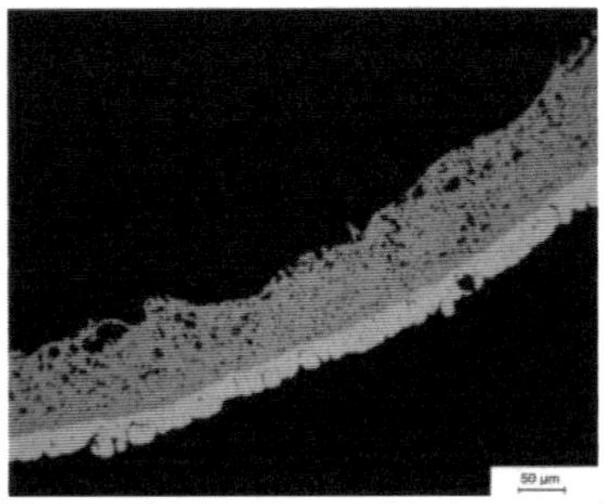

Figure 6. Metallographic image of MHS wall with copper coating.

The conducted electric sintering at free filling of MHS allowed consolidation of spheres and formation of a rigid structure (Figure 7). However, a negative aspect is the volatility of properties of the resulting sintered body. This was caused by sintering of MHS to the surface of electrodes and sometimes by the lack of contact between the spheres.

The total porosity of the MHS structure is determined by macroporosity and microporosity. Macroporosity is the space between spheres and a hollow volume inside spheres. Microporisity is tiny pores in the spheres wall and coating. Macroporosity accounts for about 90% of the total volume. It is determined by the spatial arrangement of spheres, where the highest porosity is achieved at geometrically regular square packing of the spheres. The combined regime of pressing and sintering allowed creation of better compaction directly during the sintering process. To create more complicated and purposefully designed spatial structures of MHS for further sintering, additional tools can be used allowing specific orientation or location of spheres in a magnetic field (Figure 8). The method of orientation of MHS in a permanent magnetic field is based on ferromagnetic properties of MHS [11]. A kind of the method realization is usage of a perforated steel tape 1 mm thick with a desired configuration of open space for MHS, where the steel tape takes the function of the form and the magnetic guide.

Figure 7. General view of sintered MHS sample.

Figure 8. Orientation of MHS in magnetic field.

The purposes of pilot ultrasonic trials on MHS structures were testing of possibilities of registration of ultrasonic waves through the structure and evaluation of the quality of the consolidation between the spheres. MHS can be considered at different structural levels in relation to ultrasonic wavelengths. At high frequencies in the megahertz range, where the wavelength is lesser than the sphere's size, ultrasonic wave can propagate along a complex trajectory passing along the walls and surfaces of contact between of spheres consolidated by sintering. At low and medium frequencies, from tens of kilohertz to hundreds of kilohertz, where the wavelength is much greater than the spheres' diameter, the ultrasound carrier is the entire structure, which can be considered as a highly porous but homogeneous material.

MHS structures were tested by through transmission using a pair of broadband ultrasonic piezoelectric transducers with work frequencies of 100, 300, and 1500 kHz. Excitation was produced by sine tone-bursts of 2 periods at the selected frequencies. Spectral characteristics of the received signals were assessed by FFT. Tests showed that due to the high porosity of MHS structures, only waves in a low ultrasonic frequency range can propagate. In particular MHS structures with spheres diameter from 3 to 5 mm, the upper frequency of the detection range was limited by 300 kHz. At frequencies of 1 MHz and higher, no signals were observed. The frequency dependence of ultrasound propagation has changed significantly, when the tests were carried out in the water-filled MHS structures. Because of high open porosity and the presence of large spaces between the spheres, MHS structure was easily filled with water. Three-phase (metal-air-water) medium was formed, where the continuous volumetric liquid component volume appeared conducting sound also at high frequencies. Thus the high frequency components of the spectral characteristics of ultrasound conductivity appeared (Figure 9). The observed effect can be used to evaluate the open porosity by comparing of the spectral characteristics of the structure in dry and liquid-filled conditions.

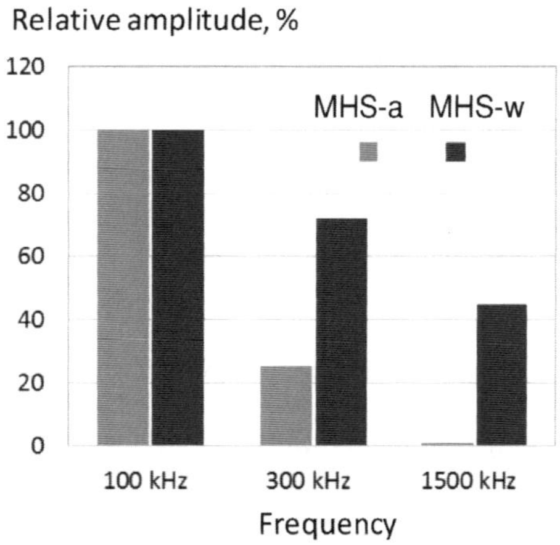

Figure 9. Relative amplitude of ultrasonic signals at frequencies of 100, 300 and 1500 kHz in percentage of the same at 100 kHz assumed as 100% in sintered MHS in air (MHS-a) and in water-filled condition (MHS-w).

To test the sensitivity of ultrasound propagation parameters to sintering quality and strength (stiffness) of bonds between spheres, sintering temperature was chosen as a variable. MHS with a copper coating 15-30 micron were used. With the increase of sintering temperature and approaching it to the melting point of copper, the contact surface area between MHS and the diffusion between layers in neighbouring spheres increased, thereby increasing strength of sintering. Ultrasound velocity was measured by through transmission at 100 kHz referring to the arrival time of the ultrasonic signal. By increasing the sintering temperature from 900 to 1000 ^{0}C, ultrasound velocity increased significantly - from 1850 to 2620 m/s (Figure 10) demonstrating sensitivity of ultrasound to the quality of sintering.

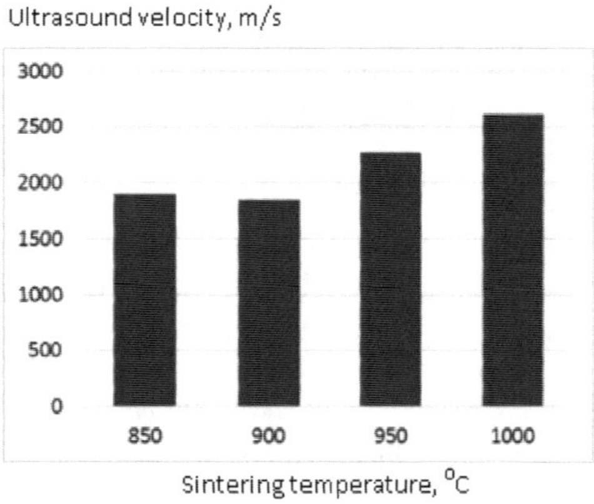

Figure 10. Ultrasound velocity in MHS structures depending on sintering temperature.

The absolute values of ultrasound velocity in the sintered MHS structure was much lower than in steel material that is explained by the hollow volume of MHS and total high porosity. Thus, additional accounting for structural parameters are is required to quantify the quality of sintering in absolute terms.

References

[1] Degischer H and Kriszt B 2002 Handbook of Cellular Metals ed H-P Degischer and B Kriszt (Weinheim, FRG: Wiley-VCH Verlag GmbH & Co. KGaA)
[2] Petrov M, Bast Y, Petrov P and Sheypak A 2012 Modern methods of producing hollow metal microspheres (in Russian; Sovremennye metody polyh metallicheskih mikrosfer/ Современные методы получения полых металлических микросфер) Trans. Moscow State Tech. Univ. MAMI 2 7
[3] Miyoshi R, Yoshimura H and Mihara Y 2010 Consolidating of Metallic Hollow Spheres Structure by Electrical Heating Int. Conf. Celluar Mater. 27–33
[4] Andersen O, Waag U, Schneider L, Stephani G and Kieback B 2000 Novel Metallic Hollow Sphere Structures Adv. Eng. Mater. 2 192–195
[5] Mironov V and Wolf H 1979 Verdichten von MetallpulverndurchelektromagnetischeKräfte 5. III Wiss.Konf.der IHZ (Zwickau) pp 128–129
[6] Mironov VA 1980 Magnetic consolidation of powder materials /in Russian; Magnitno-impulsnoe pressovanie poroshkov/ Магнитно-импульсное прессование порошков. (Riga, Zinatne)
[7] Mironov V, Zemchenkov V and Lapkovsky V 2009 Model of the Magnetic packing of dispersed materials Trans. VŠB- Tech. Univ. Ostrava 249–53
[8] Queheillalt DT, Sypeck DJ and Wadley HNG 2002 Ultrasonic characterization of cellular metal structures Mater. Sci. Eng. A~323 138–147
[9] Castaings M, Siryabe E, Renier M, Meziane A and Galy J 2015 Ultrasonic characterization of cohesive and adhesive properties of adhesive bonds J. Acoust. Soc. Am. 138 1766–1766
[10] Shishkin A, Kozlov V, Drozdova M, Rulov J, Mironov V and Hussinova I 2016 Low-density Composite Powder for Lightweight Parts Made By Means Powder Metallurgy Proc. of World Congress of Powder Metallurgy p (in print)
[11] Mironovs V, Shishkin A and Zemcenkovs V Composite material with hollow spheres, a method and device for manufacturing there of patent application. EP13197578.1. //RTU Nr. 23, Nr. LAP 2/2451

Strength analysis of small-sized thin ceramic discs obtained by spark-plasma sintering

E V Nefedova[1], V I Goltsev[1], E G Grigoryev[1], A G Zholnin[1] and K L Smirnov[2]

[1]National Research Nuclear University MEPhI, Kashirskoe shosse, 31, Moscow, 115409, Russia

[2]Institute of Structural Macrokinetics and Materials Science Russian Academy of Sciences, Akademika Osipyana str. 8, Chernogolovka Moscow region, 142432, Russia

E-mail: eugengrig@mail.ru

Abstract. The study develops an original method of mechanical test of thin small-size disks for determining resistance of brittle materials to rupture. Based on this method a test of ultimate flexture strenght was conducted on Al_2O_3 ceramics and composites of β–Si_5AlON_7–BN, system obtained by the method of spark-plasma sintering. The obtained test results allowed determining the optimal process parameters for sintering this ceramic materials in terms of their strength level: composition and dispersion of the starting powder mixtures, the maximum sintering temperature, etc.

1. Introduction

Development of advanced technology of material consolidation using electromagnetic fields requires development of appropriate mechanical test methods for determination of the mechanical properties of the produced materials. This is also conditioned by the fact that in most cases the produced compacts have a very small size with diameter 10 mm and thickness 1 to 10 mm. Known standard test methods are not applicable to the samples so small. Consequently, the objectives of this work were to develop the methods for determining the strength of small-sized samples basing on the solution of the flexible small-size disk bending problem, and to test the applicability of the developed bending test method for ceramic disk samples obtained by spark-plasma sintering (SPS).

2. Method of determination of the bending strength of small-size disk brittle material

The scheme applied to determine the strength of the material by testing small-size thin discs placed on a ring bearing is presented in Figure. 1.

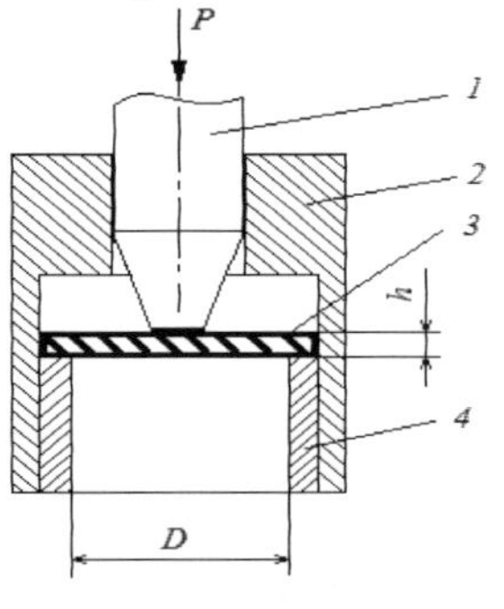

Figure 1. Flexure test of the disk supported along the contour: 1 – punch, 2 – cage, 3 – sample, 4 – bearing ring.

This test scheme was previously designed to determine the bending strength of brittle materials for larger samples (about 50mm to 60 mm in diameter and thickness of 3 mm to 6 mm) [1]. To substantiate its applicability to small-sized samples, the elastic bending disk problem has been solved, which was supported along the contour and loaded in the center with a distributed load. A scaling ratio was maintained between the geometric dimensions of large and small-size samples. A solution was obtained for circumferential and radial stresses in the central area of the disc, which coincides with the known, and the allowed relation between the thickness of the disk and the diameter of the bearing surface was defined. The calculation of the failure stress takes into account the maximum load withstood by the sample before the failure, with a linear diagram of load P - deflection ω by the formula (1):

$$\sigma_\theta = \sigma_r = \sigma_{max} = \frac{3P_{max}}{8\pi h^2}\left[4-(1-\mu)\left(\frac{d}{D}\right)^2 + 4(1+\mu)\ln\frac{D}{d}\right] \quad (1)$$

where h – the thickness of the disk; d and D – diameters of the punch and the bearing respectively; μ – Poisson's ratio.

Conventional elastic modulus was determined by converting the elastic deflection of the disk center into the circumferential and radial deformations by the formula (2):

$$\varepsilon_\theta = \varepsilon_r = \frac{h}{2R} \quad (2)$$

where R – curvature radius of the neutral plane of the disk as shown in Figure. 2.

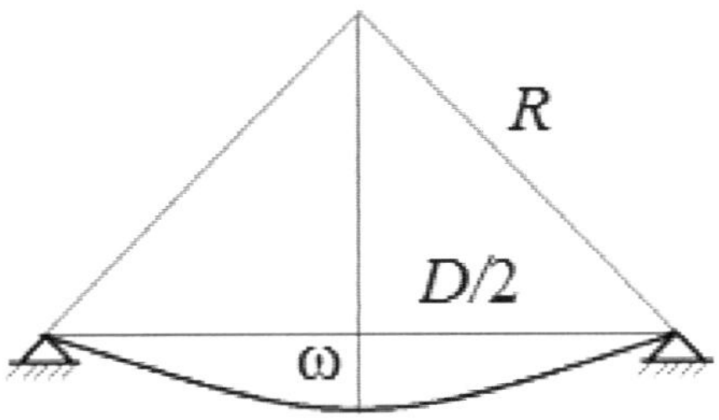

Figure 2. Calculation scheme of curvature radius of the neutral plane of the disk

The radius R is found by simple geometric proportions and with a sufficient accuracy by the formula (3):

$$R = \frac{D^2}{4\omega} \quad (3)$$

After the substitution of the formula (3) in (2), a formula for the relative deformations is obtained:

$$\varepsilon_\theta = \varepsilon_r = \frac{2h\omega}{D^2} \quad (4)$$

By Hooke's law, the deformations for the biaxial stress state are equal:

$$\varepsilon_\theta = \frac{1}{E}[\sigma_\theta - \mu\sigma_r] = \frac{\sigma_{max}}{E}(1-\mu) \quad (5)$$

IOP Conf. Series: Materials Science and Engineering **218** (2017) 012010 doi:10.1088/1757-899X/218/1/012010

By equating the (4) and the (5), a formula for determining the conventional elastic modulus is obtained:

$$E = \frac{(1-\mu)\sigma_{\max} D^2}{2h\omega} \tag{6}$$

The conventionality of *E* characteristic is due to a number of assumptions made in the derivation the formula. Conventional elastic modulus is essentially a characteristic of flexural rigidity of the disk.

3. Bending test results for small-size disks

The developed method is used to determine the strength characteristics of the ceramic composites of the β–Si_5AlON_7–BN system, obtained by SPS of the mixtures of corresponding powders. These composites consist of solid SiAlON matrix with high elastic modulus and inclusions of low-modulus particles of hexagonal boron nitride. This structure is able to provide an improved resistance to various influences by absorbing and dispersing the elastic energy released during the expansion of cracks.

Another important advantage of ceramic composites of this type is their good processability by traditional instruments. By varying the maximum temperature of isothermal holding during the SPS from 1550 °C to 1750 °C and the composition of the starting powder mixture, the samples were obtained with a relative density of 73% to 99% and h-BN content of 0 wt%. to 30 wt%. The samples were disk-shaped with 10 mm diameter and thickness of 1mm to 1.3mm. During the flexture test the bearing ring had a diameter of 7.5 mm.

The performed tests showed that the failure these samples has a pronounced brittle nature with the formation of fission fragments. The key parameter for determining the level of the bending strength of the composite is its relative density (Fig. 3).

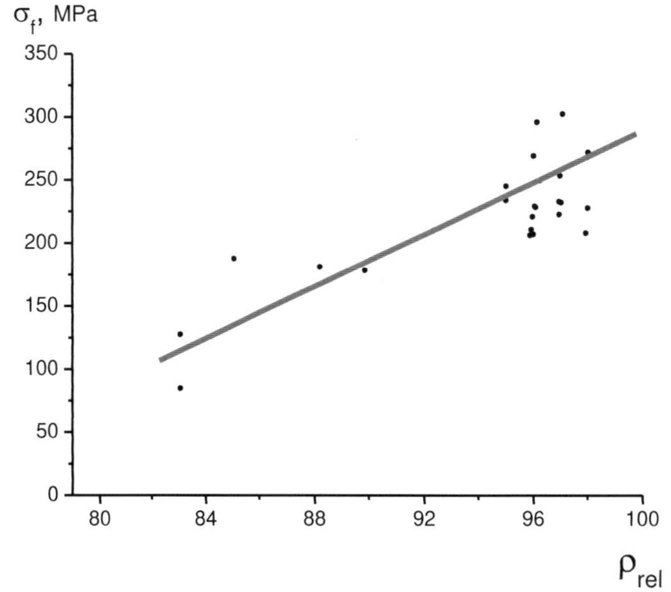

Figure 3. The dependence of bending strength of the composite β-Si_5AlON_7-BN on its relative density

It should be noted that determined level of strength characteristics of the studied samples corresponds to the strength characteristics and similar dependencies of ceramic composites with similar composition obtained by other methods[2-4].

It should also be noted that in all mentioned test the bending strength measurement was carried out by the conventional three-point bending method using a standard size samples. It was also determined that varying the h-BN content in the studied composites in the range of 0 wt % to 30 wt % has almost no effect on its strength characteristics (Fig. 4). Concurrently, a notable increase in the strength values of 30% to 40% can be achieved by substituting 20 wt% to 40 wt% of SiAlON phase in the composite content by other refractory compounds with high values of elastic modulus (SiC and TiN).

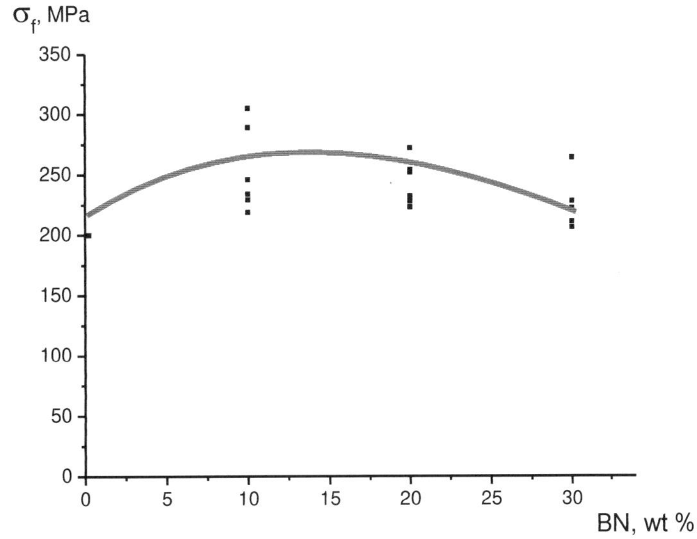

Figure 4. Dependency of the bending strength of the composite β-Si_5AlON_7-BN on BN content (ρ_{rel}=0,95-0,98).

The presented method was also used for conducting comparative mechanical tests of Al_2O_3 ceramics obtained by SPS of two kinds of nanoscale spherical powder with an average particle size of 45 nm and 150 nm and specific surface area of 35.8 m^2/g and 10.1 m^2/g respectively. The heating rate during the SPS equaled 100 °C/min, the maximum temperature of the process amounted to 1400°C, the duration of holding at the maximum temperature equaled 10 min, the diameter of sintered sample equaled 10 mm, its thickness was about 1mm to 1,3 mm. It was established that ceramics sintered from a larger size powder of Al_2O_3 has a homogeneous fine-grained structure and a high density and strength characteristics (Table 1).

Table 1. Characteristics of Al_2O_3 ceramics.

Average grain size of the starting powder (nm)	Characteristics of the sintered ceramics					
	Average grain size (μm)	Relative density (%)	Ultimate flextural strength σ_f, (MPa)	Elastic modulus E, (GPa)	Microhardness (GPa)	Structure
45	1,10	93,7	141	66	12,9	layered
150	0,65	95,8	174	73,7	19,5	homogeeous

Ceramics sintered from a finer powder of Al_2O_3 had a pronounced layered structure with higher porosity and grain size (Fig. 5).

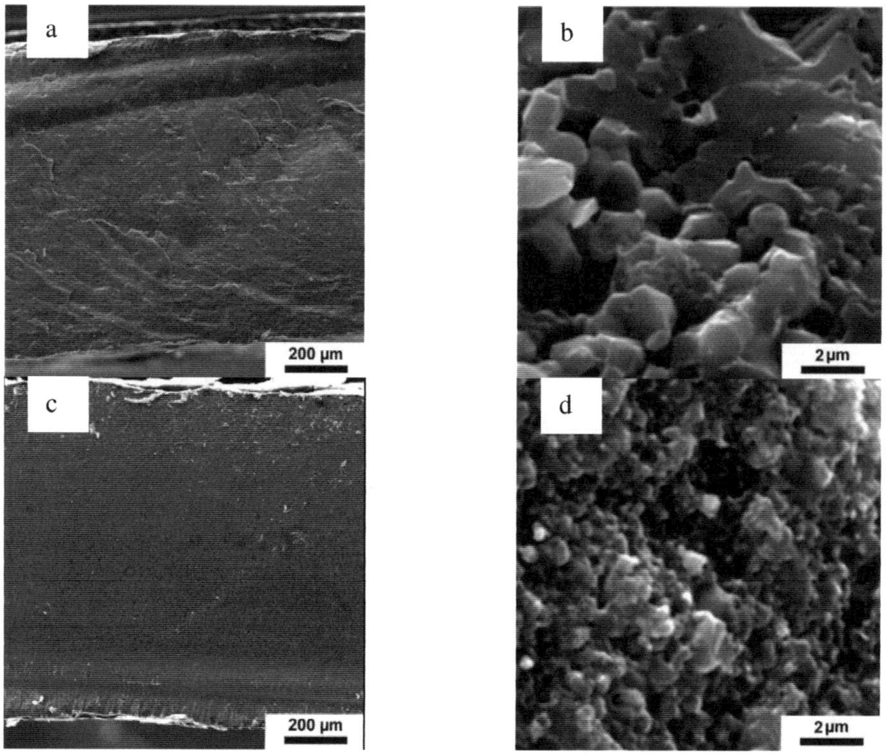

Figure 5. Fracture microstructure of Al_2O_3 ceramics sintered from powders with average particle size equal 45 nm (a,b) and 150 nm (c,d).

There are two factors negatively affecting the strength characteristics of the ceramics: large specific surface area of the starting powder, which promotes the sorption of various gases at room temperature, and the formation of large agglomerates of its fine particles. All this complicates the removal of pore in the process of sintering and leads to the formation of structural inhomogeneities, while also leads to an increase of grain size.

4. Conclusions

The research analyzed the application possibility of the developed method for determining the bending strength of ceramic materials obtained by SPS with mechanical test of small-size thin disks placed on a ring bearing.

Acknowledgments

The work was carried out with the financial support of the Russian Science Foundation (grant № 16-19-10213). The Interdepartmental laboratory of perspective technologies of creation of new materials at the National Research Nuclear University MEPhI was established under the program of mega-grants of the Russian Government (grant № 11.G34.31.0051).

References

[1] Tumanov A T 1974 Methods of test, control and investigation of machine-building materials *Mashinostroenie [Machine-building]* **2** 320
[2] Yuan B *et al* 2009 *Ceramics International* **35** 2155
[3] Zhang G-J *et al Key Eng. Mater.* **237** 123
[4] Smirnov K L *et al* 2015 *International Journal of Self-Propagating High-Temperature Synthesis* **24 (4)** 219

Studies into the impact of mechanical activation on optimal sintering temperature of UFG heavy tungsten alloys

A V Nokhrin[1], V N Chuvil'deev[1], M S Boldin[1], N V Sakharov[1], G V Baranov[2], A A Popov[1], E A Lantcev[1], V Yu Belov[2] and E S Smirnova[1]

[1] Lobachevsky State University of Nizhny Novgorod, 23 Gagarin Ave., Nizhny Novgorod, 603950, Russia.
[2] Russian Federal Nuclear Center All-Russian Research Institute of Experimental Physics, 37 Mira Ave., Sarov, Nizhny Novgorod region, 607188, Russia

E-mail: nokhrin@nifti.unn.ru

Abstract. The paper dwells on the research conducted into sintering mechanisms, the structure and mechanical properties of ultrafine-grained heavy tungsten W-Ni-Fe alloys. The dependence of alloy density on temperature of sintering (T_{sint}) is found to be nonmonotonic with a maximum equivalent to the optimal sintering temperature. Studies also encompassed the impact that the size of tungsten particles may have on the optimal T_{sint}. An increase in time of mechanical activation (MA) and acceleration of grinding bodies accompanied by a decrease in alloy particle size and formation of non-equilibrium solid solutions is shown to reduce the optimal T_{sint} of alloys. High-energy MA and Spark Plasma Sintering methods were applied to obtain samples of tungsten alloys with high mechanical properties: macroelasticity limit of up to 2,250 MPa, yield strength of up to 2,500 MPa.

1. Introduction

Heavy tungsten alloys are materials with limited mutual solubility that are of great interest both in terms of their applications and as unique objects to study the effect of interfacial and grain boundaries on physico-chemical and physico-mechanical properties of materials.

Currently, the most common method used to produce sizeable pieces from W-Ni-Fe alloys is sintering in hydrogen (HS) at temperatures higher than the melting temperature of the most fusible alloy phase - liquid phase sintering [1, 2].

Tungsten alloys obtained through liquid phase sintering have a coarse-grained heterogeneous structure and, consequently, low mechanical properties: yield strength of sintered W-Ni-Fe alloys with tungsten content of 90-95 wt% does not exceed 1,000 MPa [3, 4]. This factor significantly limits applicability of these materials in construction elements which shall meet high strength requirements.

In this context, of utmost interest is solid-phase sintering of refractory tungsten-based alloys and improved mechanical properties due to the formation of ultrafine-grained (UFG) structures.

The Spark Plasma Sintering (SPS) is considered an advanced method applied to obtained materials with high-density UFG structures [5-8].

The goal of the paper is to study the mechanisms of sintering high-strength W-Ni-Fe alloys and particularly focus on the impact that the size of tungsten particles and regimes of mechanical activation (MA) have on the optimal temperature of sintering of heavy tungsten alloys.

2. Materials. Experimental techniques

Alloys were produced using tungsten powder with average particle size of 3.9 μm, nickel powder with average particle size under 20 μm, iron powder with average particle size under 11 μm and cobalt powder with particle size under 45 μm.

Sintering powders were prepared by adding to the original coarse-grained formula of ultrafine W-Ni-Fe powders obtained though DC arc plasma synthesis (samples of series No.1), low-energy MA (samples of series No.2) and high-energy MA (samples of series No.3, 4) of coarse-grained W-Ni-Fe powders compositions (see Table 1).

The obtained powders were sintered using the method of sintering in hydrogen (HS) (series No.1-3) and Spark Plasma Sintering (SPS) (series No.4).

Low-energy MA (acceleration of grinding bodies equaling 0.4g, where g stands for gravitational acceleration) was performed in a ball mill MV0-10. High-energy MA was performed in a planetary mill APF-3 with acceleration of grinding bodies equaling 60g (rotation rate V_{MA}=1450 rpm).

HS of samples was performed in a standard resistance furnace in a two-stage mode: heating at 25°/min to 950°C, equalizing at this temperature for two hours, heating at V_h=5°/min to reach the sintering temperature. Powder samples of series No.4 were sintered through SPS using a sintering plant Dr.Sinter model SPS-625 in vacuum (6 Pa). Sintering temperatures varied in the range from 900°C to 1300°C. The temperature was measured with a pyrometer focused on the outside of the graphite mold. Samples were heated with current impulses: current impulse time reaching 3.3 ms. Sintering rate ranged from 50 to 300°/min. Samples were sintered under pressure of 50-70 MPa.

The structure of sintered alloys was studied with an optical microscope Leica IM DRM and scanning electron microscope Jeol JSM-6490. The density of samples was measured using the hydrostatic weighing method in distilled water using Sartorius CPA scales.

To study the mechanical properties of sintered samples, we applied the relaxation tests method to determine in compression tests such values as macroelasticity limit σ_o and yield strength σ_y [9]. The Hall-Petch coefficient (K) was calculated by formula: $K=(\sigma_y-\sigma_o)d^{-1/2}$.

Table 1. Materials and preparation modes.

No.	Alloy (wt.%)	Average size of tungsten particles (μm)		MA mode	Sintering mode
		Initial state	After MA		
1	95%W-3.5%Ni-1.0%Fe-0.5%Co	0.12-0.20	~0.1-0.15	Low-energy MA (0.4g, t_{MA}=72 h, V_{MA}=100 rpm)	HS (V_h=5 °/min, sintering temperature T_{sint}=1200-1500°C)
2	95%W-3.5%Ni-1.0%Fe-0.5%Co	3.9	~1.0	Low-energy MA (0.4g, t_{MA}=5 h, V_{MA}=46 rpm)	HS (V_h=5°/min, T_{sint}=1200-1500°C)
3	95%W-3.5%Ni-1.5%Fe	3.9	0.1	High-energy MA (60g, t_{MA}=20 min, V_{MA}=1450 rpm)	HS (V_h=5°/min, T_{sint}=1200-1500°C)
4	95%W-3.5%Ni-1.5%Fe	3.9	0.1	High-energy MA (60g, t_{MA}=20 min, V_{MA}=1450 rpm)	SPS (V_h=50-300°/min, T_{sint}=900-1300°C)

3. Experimental results

Figure 1a shows the dependence of density of 95W-3.5Ni-1.0Fe-0.5Co alloy on HS temperature. The dependence of density (ρ) of 95W-3.5Ni-1.0Fe-0.5Co alloy added with 10% of UFP mixture on sintering temperature is three-stage. The first stage sees monotonic growth in the alloy density from 15.77 to 17.77 g/cm^3 (from 87.1% to 98.2% of the theoretical density ρ_{th}) as temperature of sintering (T_{sint}) rises from 1200°C to T_1=1400°C. However when the T_{sint} reaches T_2~1450°C, there is a slight decrease in density to 17.53 g/cm^3 (to 96.85%). Further increase in T_{sint} to 1500°C ensures density growth to17.67 g/cm^3 (97.6% of the theoretical alloy density, see Table 1, Fig. 1a). Similar dependencies $\rho(T_{sint})$ were obtained for alloys added with 20% UFP, 100% UFP alloys are similar in nature.

It shall be noted that the above dependencies $\rho(T_{sint})$ differ from any dependences obtained while sintering coarse-grained powders (see Fig. 1a): dependence $\rho(T_{sint})$ for such materials is monotonic in nature with maximum density reached at T_{sint} corresponding to the melting point of γ-phase.

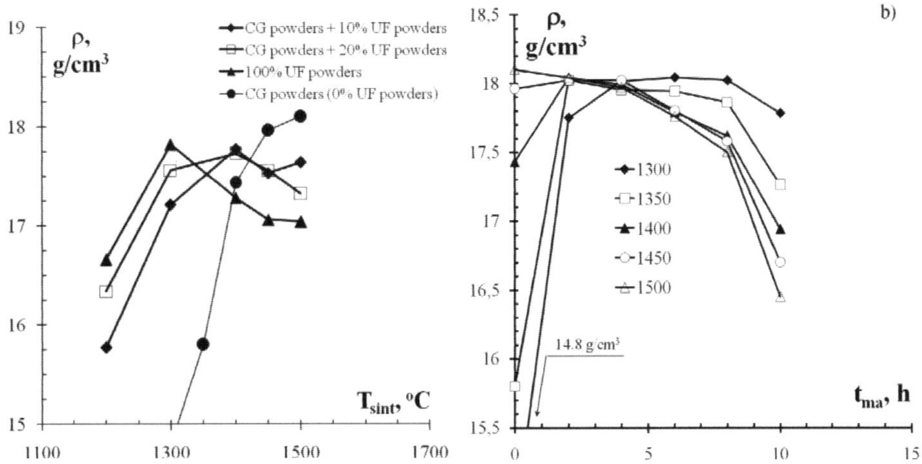

Figure 1. Dependence of density in samples of series No.1 (a) and series No.2 (b) exposed to low-energy mechanical activation on HS temperature.

It shall be also noted that the effect of density reduction at the second sintering stage (at $T_1 \leq T_{sint} \leq T_2$) is not connected with the growth of porosity in the alloy – detailed metallographic and electron microscopic studies revealed no change in the volume ratio of pores – 95W-3.5Ni-1.0Fe-0.5Co alloy samples after sintering have a high density homogeneous structure with density reaching 17.73-17.82 g/cm^3 after sintering at T_1, which approximates 98.0-98.5% of the theoretical alloy density.

Research into mechanical properties of 95W-3.5Ni-1.0Fe-0.5Co alloys in a variety of structural states shows that increase in the UFP volume ratio from 10% to 100% in alloys sintered at optimal temperature T_1 leads to greater σ_o rising from 230 MPa to 635 MPa and greater σ_y rising from 655 MPa to 905 MPa (see Table 2). Curiously enough, the Hall-Petch coefficient (K) decreases with the increase in the grain size in alloys (see Table 1).

Table 2. Impact of ultrafine powder (UFP) additives on optimal sintering temperature, density and mechanical properties of a 95W-3.5Ni-1.0Fe-0.5Co alloy (series No.1)

UFP ratio in alloys (%)	Representative sintering temperatures		Alloy density at optimal sintering temperature T_1		Average grain size	Mechanical properties of alloys		
	T_1 (°C)	T_2 (°C)	ρ (g/cm^3)	ρ (%)	d (μm)	σ_o (MPa)	σ_y (MPa)	K (MPa·m$^{1/2}$)
0	1450	1500	18.02	99.6	22	-	635	-
10	1400	1450	17.77	98.2	15	230	655	1.6
20	1350-1400	1450	17.73	98.0	10	240	665	1.3
100	1300	1450	17.82	98.5	4	635	905	0.5

The dependence of alloy series No.2 density on temperature of sintering $\rho(T_{sint})$ is found to be nonmonotonic which is similar to the dependence $\rho(T_{sint})$ for samples of series No.1 (see Fig.1b).

It should be noted that the dependence $\rho(T_{sint})$ is also influenced by mechanical activation time (t_{MA}): an increase in t_{MA} from 2 to 10 hours leads to a decrease in density of alloys sintered at T_{sint}=1500°C from 18.02 to 16.45 g/cm^3. It has been observed that as well as with samples of series

IOP Conf. Series: Materials Science and Engineering **218** (2017) 012011 doi:10.1088/1757-899X/218/1/012011

No.1, structural research revealed no increase in the volume ratio of pores at the second sintering stage at $T_1 \leq T \leq T_2$.

Table 3. Impact of low-energy MA time on optimal sintering temperature and structure of 95W-3.5Ni-1.0Fe-0.5Co alloy (series No.2)

t_{ma} (h)	Optimal temperature	Alloy density at optimal sintering temperature $T_{sint}=T_1$		Average grain size	Mechanical properties of alloys		
	T_1, (°C)	ρ (g/cm³)	ρ (%)	d (μm)	σ_o (MPa)	σ_y (MPa)	K (MPa·m$^{1/2}$)
0	1450-1500	18.08-18.10	100	50-70	-	-	-
2	1450	18.02-18.05	99.6-99.7	50	280	550	1.9
4	1400	18.00	99.5	36-38	300	550	1.5-1.6
6	1350	17.95-18.00	99.2-99.5	15-20	320	590	1.0-1.2
8	1300	17.97	99.3	5-10	740	930	0.4-0.6
10	≤1300	17.78	98.2	3-5	780	1000	0.4-0.5

The dependence of average tungsten particle size and alloy density after high-energy MA (series No.3) on sintering temperature is shown on Fig. 2a. The above data prove that an increase in alloy heating temperature above optimal sintering temperature T_1 leads to accelerated grain growth: with alloys sintered at T_{sint}=1100°C and 1450°C, the average grain size is 1.5 and 22 μm respectively, which is an order of magnitude bigger than the average size of tungsten particles in the original state.

The dependence $\rho(T_{sint})$ of a 95W-3.5Ni-1.5Fe alloy is similar to analogous dependences for alloys of series No.1 and No.2 and has two stages (Fig. 2a). As before, within the T_{sint} range from 1350°C to 1450°C corresponding to the effect of density reduction (see above), detailed electron microscopic and metallographic studies revealed no increase in the volume ratio of pores: materials sintered at T_{sint}=1350°C have a high-density structure with the volume ratio of pores not exceeding 0.2%.

Dependencies of σ_o and σ_y in a 95W-3.5Ni-1.5Fe alloy (series No.3) on temperature of sintering are two-stage; maximal mechanical properties (σ_o=1030 MPa, σ_y=1320 MPa) are ensured in an alloy sintered at T_{sint}=1250°C. Further T_{sint} increase leads to reduced material strength caused by recrystallization processes. Similar to samples of series No.1 and 2, the Hall-Petch coefficient decreases while the average grain size in alloys grows.

Figure 2c shows dependencies $\rho(T_{sint})$ of 95W-3.5Ni-1.5Fe alloy (series No.4) density obtained through high-energy MA and SPS. The presented results allow to generalize that the optimal T_{sint} for 95W-3.5Ni-1.5Fe alloy approximates T_1~1100°C (V_h=100°/min). Reduction in the heating rate causes a shift in the optimal sintering temperature to higher values: optimal temperature of sintering for SPS alloys heated at the rate of 50°/min is T_1=1200°C.

Research into the structure of samples after high-energy MA and SPS show that the structure with average grain size of ~500 nm is formed in alloys at temperature T_1=1100°C.

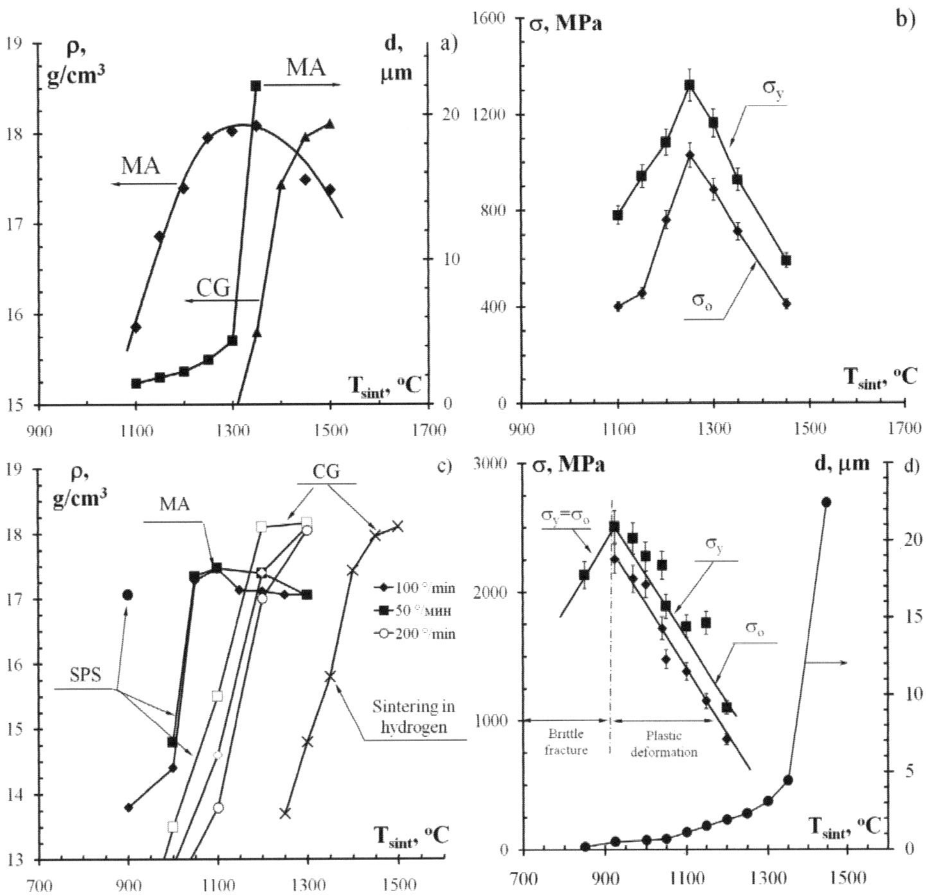

Figure 2. Dependence of density (a, c), mechanical properties and grain size (b, d) on HS temperature (a, b) and SPS temperature (c, d) of 95W-3.5Ni-1.5Fe alloy exposed to high-energy MA.

Figure 2d presents graphs showing the dependence of mechanical properties found with a 95W-3.5Ni-1.5Fe alloy (series No.4) on T_{sint} at the heating rate of 100°/min. Maximal mechanical properties of alloys (σ_o=2250 MPa, σ_y=2500 MPa) are ensured through heating at the rate of 100°/min at T_{sint}=925-950°C. However it should be noted that ductility (in compression tests) of alloys exposed to such a sintering mode is small and does not exceed 0.2-0.5%. Although SPS at higher temperatures (T_{sint}>1050°C) reduces strength properties of alloys ($\sigma_o \leq$1500 MPa, $\sigma_y \leq$ 1900 MPa), it ensures higher ductility: alloy samples are resistant to deterioration after ~10% contraction.

Values of the Hall-Petch coefficient in UFG alloys obtained using SPS appear to be much less than in fine-grained and coarse-grained alloys obtained though sintering in hydrogen (0.1-0.3 MPa·m$^{1/2}$).

4. Analysis of experimental data

4.1 Impact of MA on the dependence of density on sintering temperature

Experimental studies have shown that the dependence of samples density on the sintering temperature is three-stage in nature and has its maximum corresponding to optimal sintering temperature T_1. This result was unexpected as dependence of density on sintering temperature for common coarse-grained W-Ni-Fe alloys is monotonic.

The first rising branch of $\rho(T_{sint})$ dependence obviously reflects the accelerated process of diffusion mass transfer of tungsten in γ-phase at rising temperatures, while its behavior is rather typical of sintering of coarse-grained W-Ni-Fe mixtures. However the reason for significant density reduction – more than by 7% – at the second stage requires extra explanation.

Let us consider deformation behavior of W-Ni-Fe mixture during mechanical activation. During mechanical activation, deformation dissolution of tungsten ions in nickel-based γ-phase occurs. It shall be noted that increased MA time and acceleration of grinding bodies during mechanical activation are sure to enhance the degree of non-equilibrium in the solid tungsten solution in γ-phase.

While heating, the system tends to reduce its internal energy by means of diffusion mass transfer of excess concentration of tungsten ions from the FCC lattice of γ-phase to the particles of α-phase (tungsten), as this decreases the degree of supersaturation for nonequilibrium solid tungsten solution in nickel-based γ-phase. This process decreases density to its equilibrium value at the second stage of sintering (at $T_1<T\sim0.7T_m<T_2$) alloys (where T_m is the melting temperature of γ-phase). Further increase in alloy density at higher sintering temperatures is caused by accelerated diffusion mass transfer at elevated sintering temperature which is exponential in nature.

Thus, the reason for decreased density at the second sintering stage ($T_1<T_{sint}<T_2$) of mechanically activated tungsten alloys lies in intensive diffusion drift of tungsten atoms away from the nickel-based phase. This conclusion fully conforms to the results presented in paper [10] which showed that an increase in sintering temperature of mechanically activated 90W-7Ni3Fe alloy from 1250°C to 1400°C went along with a decrease in concentration of tungsten in γ-phase from ~50% to 36-38%.

4.2 Impact of MA on tungsten alloy strength

The analysis of results obtained from studying mechanical properties shows that the dependence of σ_y on grain size d in σ_y-$1/d^{1/2}$ coordinates may be interpolated by a straight line (see Fig. 3a). This means that for the alloys under study the following Hall-Petch relation is true: $\sigma_y=\sigma_o+K\cdot d^{-1/2}$.

It shall be noted that with an increase in grain size in sintered alloys we observe bigger K coefficient (see Fig. 3b). Minimal K values (~0.1-0.3 MPa·$m^{1/2}$) are found in alloys obtained though SPS with minimal grain growth (α-W particles grow from the initial size of ~0.1 μm to 0.95-1.1 μm). Maximal K values (~1.6-1.9 MPa·$m^{1/2}$) are observed in alloys of series No.3 that experience sizable grain growth (from the original particles size of ~1 μm to 50 μm) while sintering.

In our opinion, bigger grain boundary contribution (increased K coefficient) during grain growth results from supersaturated solid tungsten solution in γ-phase. An interfacial (α-γ)-boundary migrating at high rate 'sweeps' tungsten atoms in γ-phase. This leads to a thin layer of heavily supersaturated solid tungsten solution in γ-phase that is formed in front of a migrating (α-γ)-boundary. This layer contributes to strength (increased K value) and may partly compensate for reduced alloy strength arising from grain growth.

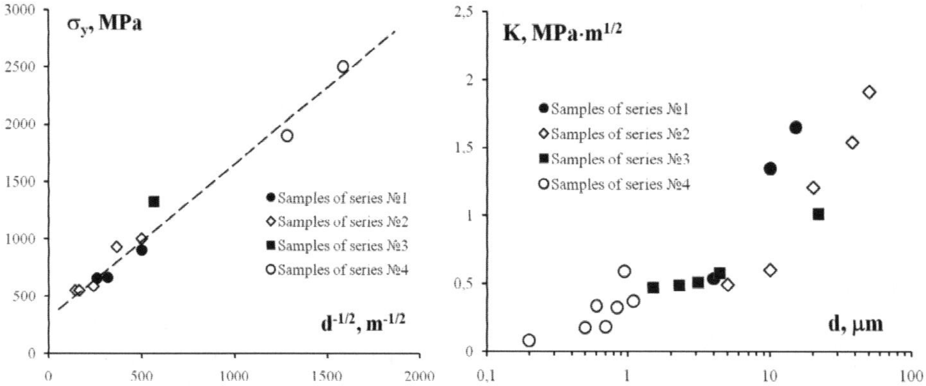

Figure 3. Dependence of yield strength σ_y on grain size (a) and dependence of Hall-Petch coefficient (K) on grain size for tungsten alloys (b)

5. Conclusion

Experimental research was conducted into the structure of mechanically activated nano- and ultrafine tungsten alloys obtained through sintering in hydrogen and SPS. The dependence of alloy density on temperature of sintering is found to be nonmonotonic with a maximum equivalent to optimal temperature of sintering. Increase in MA time and acceleration of grinding bodies during mechanical activation accompanied by decrease in alloy particle size and formation of non-equilibrium solid solutions is shown to reduce optimal temperatures of sintering of alloys.

Nonmonotonic nature of dependence of alloy density on temperature of sintering is shown to be determined by the kinetics of diffusion-controlled processes associated with the changes in tungsten concentrations in nickel-based γ-phase solid solution, as well as development of recrystallization processes at high-temperature heating of alloys.

Formation of a supersaturated solid tungsten solution in γ-phase is found to increase Hall-Petch coefficient and may ensure extra strengthening in case of intensive migration of interfacial boundaries during sintering of tungsten alloys.

Acknowledgements

The research is carried out with the support of the Grant of the President of the Russian Federation No. NSh-7179.2016.8 and Ministry of Education and Science of Russian Federation.

References

[1] Green E C, Jones D J, Pitkin W R 1954. *Symposium of Powder Metallurgy. Special Report* **(58)** 253.
[2] Krock R, Shepard H 1963 *Transaction of the Metallurgical Society of AIME* **227** **(5)** 1127.
[3] Gong X, Fan J L, Ding F, Song M, Huang B Y, Tian J M 2011 *Materials Science and Engineering A* **528** 3646.
[4] Das J, Kiran U R, Chakraborty A, Prasad N E 2009 *Int. Journal of Refractory Metals and Hard Materials* **27** 577.
[5] Tokita M 2013 *Spark Plasma Sintering (SPS) Method, Systems, and Applications (Chapter 11.2.3)*. In Handbook of Advanced Ceramics (Second Ed.). Ed. Shigeyuki Somiya, Academic Press. 1149.
[6] Orlova A I, Volgutov V Yu, Mikhailov D A, Bykov D M, Skuratov V A, Chuvil'deev V N, Nokhrin A V, Boldin M S, Sakharov N V 2014 *Journal of Nuclear Materials* **446** **(1-3)** 232.
[7] Chuvildeev V N, Panov D V, Boldin M S, Nokhrin A V, Blagoveshensky Yu V, Sakharov N V, Shotin S V, Kotkov D N 2015 *Acta Astronautica* **109** 172.
[8] Chuvil'deev V N, Boldin M S, Dyatlova Ya G, Rumyantsev V I, Ordan'yan S S 2015 *Inorganic Materials* **51** **(10)** 1047.
[9] Nokhrin A V 2012 *Deformation and Fracture of Materials* **(11)** 23 (in Russian).
[10] Fan J, Huang B, Qu X, Zou Z 2001 *Int. Journal of Refractory Metals and Hard Materials*. **19** 73.

IOP Conf. Series: Materials Science and Engineering **218** (2017) 012012 doi:10.1088/1757-899X/218/1/012012

Spark Plasma Sintering of high-strength ultrafine-grained tungsten carbide

A V Nokhrin[1], V N Chuvil'deev[1], Yu V Blagoveshchenskiy[1], M S Boldin[1], N V Sakharov[1], N V Isaeva[2], A A Popov[1], E A Lantcev[1], O A Belkin[1] and E S Smirnova[1]

[1] Lobachevsky State University of Nizhny Novgorod, 23 Gagarin Ave., Nizhny Novgorod, 603950, Russia.
[2] A.A. Baykov Institute of Metallurgy and Material Science of RAS, 49 Leninsky Ave., Moscow, 119991, Russia

E-mail: nokhrin@nifti.unn.ru

Abstract. The paper dwells on the research conducted into high-rate consolidation of pure tungsten carbide nanopowders using the Spark Plasma Sintering. Studies included the effect that the original size of WC nanoparticles and their preparation modes have on density, structure parameters, and mechanical properties of tungsten carbide. It has been found that materials that show abnormal grain growth during sintering have lower values of sintering activation energy as compared to materials the structure of which is more stable during high-rate heating. A qualitative model is proposed that explains this effect through the dependence of the grain boundary diffusion coefficient on the grain boundary migration rate.

1. Introduction

Pure tungsten carbide is of interest for a variety of applications (cutting tools, drawing dies, etc.) due to a good combination of physical and mechanical properties (high melting temperature, high hardness, low friction coefficient, and chemical resistance to corrosion and oxidation). However, high brittleness of tungsten carbide traditionally obtained with the help of a powder sintering technology does not allow to use it in its purest form. That is why the issue of achieving high mechanical properties with pure tungsten carbide is crucial.

Improving the mechanical properties of tungsten carbide is primarily associated with obtaining ultrafine grains (UFG) in the sintered material [1-2]. To this end, WC nanopowders are used as a initial material. DC arc plasma synthesis is the most advanced method for obtained of WC nanopowders [3, 4].

The Spark Plasma Sintering (SPS) is considered an advanced method applied to obtain materials with high-density UFG structure [5-8]. The core of the SPS lies in the high-rate heating of powder materials in vacuum or inert atmosphere by passing direct pulse large currents through the tooling and samples and simultaneous pressure application. High heating rates become particularly relevant during sintering of UFG materials as they limit the growth of nano- and ultra-dispersed powders and contribute to the formation of a uniform high-density UFG structure at low consolidation temperatures [6-8].

IOP Conf. Series: Materials Science and Engineering **218** (2017) 012012 doi:10.1088/1757-899X/218/1/012012

The goal of this paper is to study the structure and properties of tungsten carbide obtained through SPS from nanopowders produced using DC arc plasma synthesis.

2. Materials. Experimental techniques

Initial materials include tungsten carbide nanopowders produced using DC arc plasma synthesis [3, 4]. The second stage of obtaining α-WC nanoparticles is recovery annealing in hydrogen. Its goal is to control the volume fraction of W_2C, β-WC (WC_{1-x}), W, α-WC particles and free carbon C in a nanopowder charge based on tungsten monocarbide α-WC.

To study the impact that the particle parameters have on the structure and properties of tungsten carbide ceramics, 6 series of powders different in composition were obtained through DC arc plasma synthesis with various average initial particle size (R_0). For the purposes of comparison, this paper involves research into commercial powder α-WC produced by H.C. Starck Company (see Table 1).

Table 1. Impact of the initial particle size and volume fraction of tungsten monocarbide (α-WC) on SPS temperatures and structural parameters of sintered samples (V=25 °C/min, pressure 75 MPa).

Series No.	S (m^2/g)	R_0 (nm)	f (%)	SPS temperatures		Average gain size(2)		Activation energy	Hardness(2)	Fracture toughness(2)
				T_1 (°C)	T_2 (°C)	d_m (μm)	d_a (μm)	Q (kT_m)	H_v (GPa)	K_{1c} (MPa·$m^{1/2}$)
1	8.34	46	83.5	1145	1420	0.1	1-3	5.2	28.0	5.1
2	6.87	55	91.7	1370-1405	1495	0.1	8	5.0	26.0	5.4
3	6.01	63	93.6	1460	1520	0.3	20	4.7	20.0	5.6
4	3.36	113	99.5	1330	1380-1390	2.5-10	17-20	4.1	16.5	6.0
5	5.03	80	100	1425	1505	0.35	1.2-1.3	10.4	34.0	4.3
6	5.27	72	100	1365-1375	1490	0.1-0.15	(3)	10.8	29.0	4.2
7(1)	3.40	112	100	1430	1515	0.2-0.3	(3)	7.7	31.0	4.0

(1) – H.C. Starck powder; (2) – after sintering at 1550°C; (3) – no abnormal grain growth

According to X-ray diffraction analysis, these nanopowders, except for α-WC nanopowders, contained a different number of the following phases: W_2C, β-WC (WC_{1-x}), α-W. As seen from X-ray diffraction patterns for powders with reduced tungsten monocarbide volume fraction (f=83.5-93.6%) clearly show peaks of the primary α-WC phase, as well as β-WC phase (W_xC_{1-x}). α-W tungsten particles in the amount of ~0.5 wt% are identified in powders of Series No.4 at the background level. Sintering of nanopowders was performed using a sintering plant "Dr. Sinter model SPS-625" in the temperature range from 1400°C to 1950°C without holding. The heating rate (V) ranged from 25 to 2400°C/min. Experiments were carried out at uniaxial compression pressure (P) of 60 and 75 MPa. Sintering took place in 4 Pa vacuum in graphite molds 12 mm in inside diameter.

A dilatometer which is part of a unit "Dr. Sinter model SPS-625" was used to measure effective shrinkage (L_{eff}, mm) under continuous heating (see Fig.1a). To take account of thermal expansion of the "machine-sample" system into effective shrinkage L_{eff} measured with a dilatometer, additional studies into thermal expansion of the system were carried out without samples (empty mold). The above dependencies $L_0(T)$ were then deduced from $L_{eff}(T)$ dependence, and the dependence of shrinkage rate D (Fig. 1b) on the heating temperature was calculated.

Phase composition of samples was studied with "Shimadzu XRD-700" X-ray diffractometer. The chemical composition analysis was performed using "Ultima 2 ICP" atomic emission spectrometer. Vickers hardness (H_v) was measured using "Struers Duramin-5" microhardness tester with a 2 kg load. Fracture toughness K_{IC} was calculated using the Palmqvist method. The intercept method was applied to measure grain size using Jeol JSM-6490 scanning electron microscope.

IOP Conf. Series: Materials Science and Engineering **218** (2017) 012012 doi:10.1088/1757-899X/218/1/012012

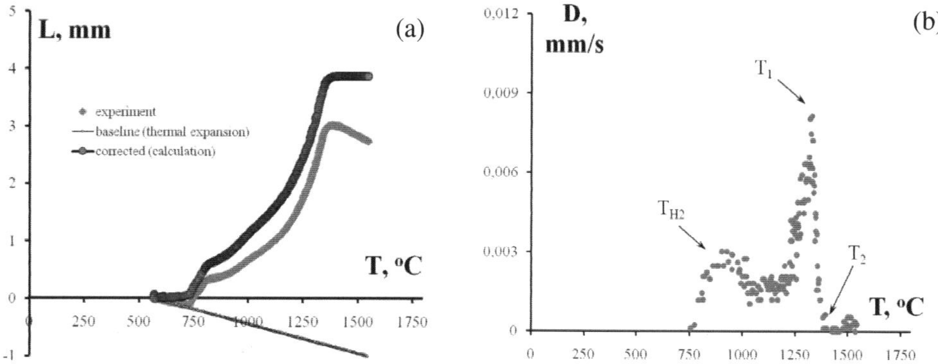

Figure 1. Dependence of shrinkage amount (a) and shrinkage rate (b) on sintering temperature of tungsten carbide nanopowders. Series No.4. Heating rate V=25 °C/min, pressure 75 MPa.

3. Experimental results

3.1. Impact of the initial particle size on tungsten carbide sintering temperature

As seen from Fig.1, D(T) dependence for all the nanopowders under consideration obviously has a three-stage character. Intensity of powder shrinkage is little at an early heating stage. At temperatures below representative temperature T_1, shrinkage rate is growing; with temperature increasing ($T_1<T<T_2$) D(T) value declines rapidly, and at $T \sim T_2$ shrinkage of sintered powders stops.

Data analysis presented in Table 1 proves that representative temperatures T_1 and T_2 depend on the initial particle size: with compositions characterized by low α-WC monocarbide volume fraction (f=83.5-93.6%) increase in the initial size of particles R_0 from 46 nm to 63 nm results in T_1 growing from 1145 °C to 1460 °C and T_2 growing from 1420 °C to 1520 °C. For powders with initial particle size R_0=113 nm and f=99.5%, temperature values are T_1=1330°C and T_2=1380-1390°C (see Table 1).

Thus, it may be claimed that should SPS nanopowders be added up with second-phase particles, nonmonotonic dependence of representative sintering temperatures T_1 and T_2 on the initial particle size R_0 is observed (see Table 1).

During sintering of nanopowders of Series 5-7 (volume fraction for tungsten monocarbide close to 100%), increase in the initial particle size R_0 from 72 nm to 112 nm leads to a monotonic increase in T_1 from 1365-1375°C to 1430°C and insignificant rise in T_2 from 1490°C to 1515°C.

3.2. Impact of nanopowder composition on parameters of the structure formed

Electron microscopic studies prove that abnormal grain growth is observed in the structure of the material sintered from nanopowders added up with second-phase particles (volume fraction of tungsten monocarbide f=83.5-99.5%). After heating to T=1460-1550°C, an inhomogeneous consertal structure is formed with average size of abnormal grains d_a (from 3 to 20 μm) being an order of magnitude bigger than the average size of grains in a matrix d_m (~0.1-0.3 μm) (see Table 1). It is notable that in case of abnormal growth, pores are mainly found along the boundaries of abnormally coarse tungsten carbide grains. With sintering temperature rising to 1700°C, a uniform coarse-grained structure is formed. In this case pores are found both along the boundaries and in the grain boundaries of the sintered material. Analysis of the data presented in Table 1 shows that with tungsten monocarbide volume fraction rising from 83.5 to 99.5%, the average size of abnormally coarse grains d_a is growing from 1-3 to 17-20 μm.

X-ray diffraction analysis illustrates that abnormally coarse grains are W_2C particles. The volume fraction of W_2C particles in sintered ceramics ranges from 2.4 to 10.7% depending on the initial volume fraction of tungsten monocarbide and sintering temperature. The volume fraction of W_2C particles (f_{W2C}) in ceramic samples of Series No.3 (f = 93.6%) and Series No. 4 (f = 99.5%) approximates ~24.4% and ~5.2%, respectively (sintering temperature 1500°C, heating rate V=25°C/min).

For ceramics of Series No.5 obtained through sintering of nanopowders, the volume fraction of tungsten monocarbide in which according to X-ray diffraction analysis constitutes 100%, W_2C particles are identified in the amount ~9.9% (sintering temperature 1500°C, heating rate V=25°C/min). This result brings about the conclusion that the initial volume fraction of the second-phase particles acting as sort of "nuclei's" ("points of grain growth") for abnormally coarse grains is small and lies within the experimental error range permissible for X-ray diffraction analysis.

Note should be taken of no correlation between the volume fraction of nonstoichiometric-phase particles (1-f) and volume fraction of newly formed W_2C particles: in ceramic samples of Series No.3 (f=93.6%), Series No.4 (f=99.5%), and Series No.5 (f~100%) sintered at 1500°C, the volume fraction of W_2C particles is f_{W2C} = 24.4%, 5.2%, and 9.9% respectively. Similar results were achieved while studying ceramics sintered at other temperatures and heating rates. In our opinion the above proves that not all nonstoichiometric-phase particles (W_2C, β-WC (WC_{1-x}), α-W) are "nuclei's" for abnormally coarse W_2C grains while sintering α-WC ceramics. (The nature of "nuclei's" for abnormally coarse grains will be discussed below.)

Research into the structure of samples sintered from nanopowders of Series No.6-7 (volume fracture of tungsten monocarbide is 100%) shows that these materials are characterized by normal grain growth and monomodal grain size distribution, although the distribution bell is rather wide.
The average grain size for samples sintered at 1400 and 1800°C is 90 nm and 350 nm respectively. All the materials under consideration are characterized by active grain growth at temperatures above 1800°C which led to the formation of an UFG structure (average grain size is 300-400 nm).

4. Analysis of experimental data

4.1 Calculation of sintering activation energy

As noted above, dependencies of the degree of shrinkage (L, mm) and shrinkage rate (D, mm/s) on temperature and heating time are considered primary data during SPS of tungsten carbide. The procedure of converting L(T) dependence into compression schedules ρ/ρ_{th}-T is specified in [9, 10] (ρ_{th}=15.77 g/cm^3 being a theoretical density of tungsten monocarbide). $\rho/\rho_{th}(T)$ dependencies have a three-stage nature typical for compression curves during sintering of powder materials. Our further analysis of $\rho/\rho_{th}(T)$ dependencies shall be limited to considering the stage of intensive compression (temperature range $T_1<T<T_2$ shown on Fig.1b).

Activation energy Q of the compression process may be determined by the slope angle of $\rho/\rho_{th}(T)$ dependence (presented in double logarithmic coordinates) on reciprocal temperature T_m/T: $\ln(\ln[(\rho/\rho_{th})/(1-\rho/\rho_{th})]) - T_m/T$ (for more detail see [9, 10]). (T_m=3143 K is the melting temperature for tungsten monocarbide).

Table 1 shows the values of the sintering activation energy Q calculated for compositions with various volume fracture of monocarbide α-WC. In samples of Series No.1-4, the volume fraction of α-WC in which is under 100%, the values of the sintering activation energy within the sintering temperature range T= 1150-1550 °C is Q=107-135 kJ/mol (~4.1-5.2 kT_m), which agrees well with the data obtained in [2] within the same temperature range of 1050-1200 °C during SPS of tungsten carbide nanopowders amid intensive grain boundary (GB) migration: Q =103.5 kJ/mol (~4 kT_m).

Note shall be taken that the obtained values of Q are much lower than the values of the activation energy for GB diffusion Q_b of carbon ^{14}C in coarse-grained tungsten carbide obtained through hot pressing (Q_b=367 kJ/mol ~14 kT_m [11]), as well as the values of the activation energy for ^{14}C diffusion provided in [12] (Q_b=300 kJ/mol ~11.4 kT_m) и в [13] (Q_b=240 kJ/mol ~9.1 kT_m).

In samples of Series No.6-7 (100% α-WC), average values of the Q appear to be considerably higher (7.7-10.8 kT_m~ 201-282 kJ/mol) and better comply with the above reported value [11-13].

Thus, data provided in Table 1 show that two types of behavior are observed during WC sintering: sintering under abnormally low activation energy values and sintering under activation energy corresponding to the equilibrium values of the activation energy for GB diffusion Q_b.

It should be emphasized that after sintering materials of all the above types significantly differ in structure: samples of type 1 materials (sample of Series No.1-4) are characterized by an inhomogeneous structure containing abnormally coarse grains, while samples of type 2 materials (samples of Series No.6-7) form a predominantly homogeneous UFG structure[1].

4.2 Impact of the average initial powder particle size and volume fraction of tungsten monocarbide on the grain growth and WC ceramics density

Let us consider the impact that nonstoichiometric-phase particles of W_2C, β-WC (W_xC_{1-x}) carbides and pure tungsten (α-W) have on sintering of tungsten carbide. Some of these nonstoichiometric-phase particles may act as sort of 'nuclei's' for abnormally coarse grains. (Further, for the sake of brevity, we shall call these particles 'active phase particles').

Let us assume that the typical size to which on average an abnormally growing 'active phase particle' may grow is determined by the distance to another such particle λ^*. Based on microstructural data, it is possible to roughly estimate the volume fraction of such 'active phase particles'. Let us take a simple formula: $\lambda^* = R\sqrt[3]{f_r}$, where f_r and R stand for volume fraction and size of 'active phase particles' respectively. Assuming that λ^* value corresponds to half the size of abnormal grains d_m ($\lambda^*=d_m/2$) and by substituting values R_0=46-113 nm and d_m=3-17 μm (see Table 1) into the equation for λ^*, we shall find that the volume faction of active phase particles' is f_r~$3\cdot10^{-5}$-$3\cdot10^{-4}$%. This value is much smaller than the values of the nonstoichiometric phase volume fraction in samples of Series No.1-4 (from 0.5 to 16.5%) measured with the help of X-ray diffraction analysis. The obtained results also count in favor of the above made assumption that not all nonstoichiometric-phase particles (W_2C, β-WC (W_xC_{1-x}), α-W) are 'active'.

It is crucial to note that the volume fraction of W_2C particles in sintered ceramics is rather high (10-25%), while the volume fraction of W_2C particles in the original powder is below the resolving power of the X-ray diffraction analysis (less than 1%). This means that the SPS process is characterized by 'transformation' of the original 'active phase particles' into W_2C particles which grow rapidly on further heating. Since the experiment shows simultaneous growth in W_2C volume fraction and size (size of abnormally coarse grains), it becomes obvious that this process is not a conventional coalescence process.

In our opinion, the most likely candidate to act as active phase particles may be particles of α-W metal tungsten in which carbon diffusion at specified sintering temperatures (1300-1500°C) is rather intense (parameters of volume carbon diffusion in monocrystals α-W: D_{v0}=3.0 cm²/s, Q_v=247 kJ/mol ~ 8 kT_m [12]), whereas α-W at elevated temperatures has a high tendency to W_2C carbide formation [14].

We shall note that W_2C phase cannot act as an 'active phase'. This is due to the fact that the volume fraction of W_2C particles in sintered ceramics is much greater than the volume fraction of W_2C particles of the initial powder. This means that the process of splitting the main phase α-WC into W_2C and α-W tungsten ($2WC \rightarrow W_2C + 2W$)[2] shall take place during the SPS process. Such methods as X-ray diffraction analysis and electron microscopy have not detected α-W in the structure of sintered ceramics.

[1] Exception to the above are ceramic samples sintered from nanopowders of Series No.5 in which W_2C particles were discovered and consequently slight abnormal grain growth was detected: the average size of abnormal grains is d_a=1.2-1.3 μm, while the average size of 'matrix' grains is d_m=0.35 μm.

[2] Chances of α-WC particles disintegration during SPS are beyond the scope of this research. Slight W_2C particle formation (under 5%) in ceramics sintered from nanopowders that entirely consist of α-WC tungsten monocarbide was detected only under such simultaneous conditions as extremely high heating rates (1500 °C/min and more) and high sintering temperatures (1720 °C and more). Since such sintering modes are beyond the considered temperature range and sintering rate range, we shall neglect the impact of this factor of the processes discussed.

The hypothesis about the role of tungsten is supported by the results of research [15] into sintering of tungsten carbide with additions of pure tungsten and carbon (pure WC, WC+0.5%W, WC+0.4%C, WC+0.5%W+0.4%C). In [15] shows that with 0.5% α-W particles, the fine structure of a sintered sample accommodates abnormally coarse grains of elongated shape, while a coarse equiaxed structure is formed by sintering other systems.

Summarizing the analysis results it may be concluded that big distance between 'active phase particles' λ^* (the role of which is presumably played by tungsten particles) leads to the fact that with the reduction in the total volume fraction of nonstoichiometric phase the size of abnormally coarse grains d_m is growing.

4.3 Effect of accelerated sintering during abnormal grain growth

Let us consider the issue related to the reasons for lowering the optimal sintering temperature and the reasons why low sintering activation energy values occur in materials with abnormal grain growth.

In our opinion, the said effects may be related to some changes in the parameters of GB diffusion due to intense grain growth in such materials.

In accordance with the concepts of the non-equilibrium GB theory [16, 17], the activation energy of GB diffusion Q_b depends on the degree of grain boundary non-equilibrium, which in turn is determined by the density of defects introduced into GB: orientational misfit dislocations (OMD) with Burgers vector Δb and density ρ_b, as well as products of their delocalization such as a tangential component of delocalized OMD with Burgers vector density w_t. As shown in [16, 18], in case of abnormal grain growth when GB migrating at high speed absorb lattice dislocations distributed in grains, the value of non-equilibrium GB diffusion D^*_b is exponentially dependent on the density of lattice dislocations and grain boundaries migration rate: $D^*_b=D_b\exp(g_1(\rho_v V_m)^n)$, where D_b stands for the diffusion coefficient in equilibrium GB, ρ_v is the density of lattice dislocations, V_m is the GB migration rate, g_1 is a numerical parameter that depends on the thermodynamic properties of the material, n=0.25, 0.5 is a numerical coefficient that depends on what types of defects dominate within grain boundaries: OMD (n=0.25) or tangential components of delocalized dislocations (n=0.5).

The results show an important practical conclusion regarding the impact of the GB migration on the process of SPS of UFG tungsten carbide ceramics. To ensure tungsten carbide sintering at a reduced temperature, it is necessary to optimize its phase composition and sintering modes so as to provide sufficiently 'fast' migration of GB while heating. However, it should be borne in mind that in case of sintering nonstoichiometric powders of tungsten carbide, the abnormal grain growth leads to a decrease in the mechanical properties of sintered samples.

Summary

1. One of the most probable causes of abnormal grain growth during SPS of tungsten carbide is the presence of α-W particles in nanopowders. At elevated temperature tungsten particles react with carbon ($2W + C \rightarrow W_2C$), which results in W_2C particles that are characterized by high migration mobility of boundaries.
2. Rapid migration of GB leads to 'sweeping' by GB of lattice dislocations and reduced activation energy of DB diffusion. This reduces the optimal sintering temperature for tungsten carbide powders. However, materials obtained at low sintering temperatures have low hardness and fracture toughness because of heterogeneity of the grain structure.
3. The effect of reducing the optimal sintering temperature for tungsten carbide powders during SPS is related to the effect of GB diffusion acceleration caused by abnormal grain growth.

Acknowledgements

This research is carried out with the support of the RFBR (Grant No. 15-33-21007) and the Ministry of Education and Science of the Russian Federation.

References

[1] Zhao J, Holland T, Unuvar C, Munir Z A 2009 *International Journal of Refractory Metals and Hard Materials* **27** 130.
[2] Nanda A K, Watabe M, Kurokawa K 2011 *Ceramics International* **37 (7)** 2643-2654.
[3] Krasovskii P V, Malinovskaya O S, Samokhin A V, Blagoveshenskiy Y V, Kazakov V A, Ashmarin A A 2015 *Applied Surface Science* **339** 46.
[4] Isaeva N V, Blagoveshchenskii Y V, Blagoveshchenskaya N V, Mel'nik Y I, Samokhin A V, Alekseev N V, Astashov A G 2014 *Russian Journal of Non-Ferrous Metals* **55 (6)** 585.
[5] Tokita M 2013 *Spark Plasma Sintering (SPS) Method, Systems, and Applications (Chapter 11.2.3)*. In Handbook of Advanced Ceramics (Second Ed.). Ed. Shigeyuki Somiya, Academic Press. 1149.
[6] Orlova A I, Volgutov V Yu, Mikhailov D A, Bykov D M, Skuratov V A, Chuvil'deev V N, Nokhrin A V, Boldin M S, Sakharov N V 2014 *Journal of Nuclear Materials* **446 (1-3)** 232.
[7] Chuvildeev V N, Panov D V, Boldin M S, Nokhrin A V, Blagoveshensky Yu V, Sakharov N V, Shotin S V, Kotkov D N 2015 *Acta Astronautica* **109** 172.
[8] Chuvil'deev V N, Nokhrin A V, Baranov G V, Moskvicheva A V, Boldin M S, Kotkov D N, Sakharov N V, Blagoveshensky Yu V, Shotin S V, Melekhin N V, Belov V Yu 2013 *Nanotechnologies in Russia* **8 (1-2)** 108.
[9] Chuvil'deev V N, Boldin M S, Dyatlova Ya G, Rumyantsev V I, Ordan'yan S S 2015 *Inorganic Materials* **51 (10)** 1047.
[10] Potanina E, Golovkina L, Orlova A, Nokhrin A, Boldin M, Sakharov N 2016 *Journal of Nuclear Materials* **473** 93.
[11] Buhsmer C P, Crayton P H 1971 *Journal of Materials Science* **6 (7)** 981.
[12] Properties, Production and Application of Refractory Compounds. Handbook / Ed. Kosolapova T.Y. 1986 (Moscow: Metallurgiya) 928 p. (in Russian)
[13] Erdelyi G., Beke D.L. Chapter 11 Dislocation and grain boundary diffusion in non-metallic system. In book "Diffusion in Non-Metallic Solids (Part 1)". 1999 (London-Bornstein – Group III Condensed Mater) **33B1** 1.
[14] Kharatyan S L, Chatilyan H A, Arakelyan L H 2008 *Materials Research Bulletin* **43 (4)** 897.
[15] Gubernat A, Rutkowski P, Grabowski G, Zientara D 2014 *International Journal of Refractory Metals and Hard Materials* **43** 193.
[16] Chuvil'deev V N Non-equilibrium grain boundaries in metals. Theory and applications. 2004 (Moscow: Fizmatlit) 304 p. (in Russian)
[17] Chuvil'deev V N, Kopylov V I, Zeiger W 2002 *Annales de Chimie: Science des Materiaux* **27(3)** 55.
[18] Nokhrin A V 2012 *Technical Physics Letters* **38 (7)** 630.

IOP Conf. Series: Materials Science and Engineering **218** (2017) 012013 doi:10.1088/1757-899X/218/1/012013

The use of Spark Plasma Sintering method for high-rate diffusion welding of high-strength UFG titanium alloys

A V Nokhrin[1], **V N Chuvil'deev**[1], **M S Boldin**[1], **A V Piskunov**[1], **N A Kozlova**[1], **M K Chegurov**[1], **A A Popov**[1], **E A Lantcev**[1], **V I Kopylov**[1] **and N Yu Tabachkova**[2]

[1] Lobachevsky State University of Nizhny Novgorod, 23 Gagarin Ave., Nizhny Novgorod, 603950, Russia.
[2] The National University of Science and Technology MISiS, 4 Leninsky Ave., Moscow, 119991, Russia

E-mail: nokhrin@nifti.unn.ru

Abstract. The article provides an example of applying the technology of spark plasma sintering (SPS) to ensure high-rate diffusion welding of high-strength ultra-fine-grained UFG titanium alloys. Weld seams produced from Ti-5Al-2V UFG titanium alloy and obtained through SPS are characterized by high density, hardness and corrosion resistance.

1. Introduction

Today ultrafine-grained (UFG) metals and alloys obtained through Equal Channel Angular Pressing (ECAP) [1] generate particular interest among researchers. This is because UFG materials are found to have a unique combination of high strength and ductility [2-4], corrosion resistance [5] and radiation resistance [6], high-rate superplasticity [2, 7-9], etc.

One of the key challenges that currently prevent industrial introduction of UFG materials is the welding issue: traditional technologies such as argon arc welding or electron beam welding accompanied by metal meltdown fail to preserve the UFG structure with high physicochemical properties and performance characteristics in a weld seam. Solid phase technologies, such as classical diffusion welding or superplastic forming, require high temperatures and prolonged hold time which result in recrystallization processes and subsequent loss of any unique properties found with UFG materials.

The technology of Spark Plasma Sintering (SPS) offers great opportunities to obtain high-strength weld seams. SPS is based on the idea of high-rate heating (up to 2500°C/min) of materials (powders, bulk materials, composites) in vacuum or inert atmosphere by means of passing millisecond high-power pulses (up to 5000 A) of direct pulse current through samples and simultaneous pressure application [10]. Such a combination enables high-density structures in metals, alloys and ceramics [11-13]. Advantages of SPS technology include an opportunity to reduce optimal sintering temperature for UFG materials [11-14]. High heating rates and an opportunity to ensure significant diffusion acceleration at reduced heating temperatures are essential to limit the process of grain growth and preservation of the original UFG structure in materials.

It shall be noted that ensuring a high-density weld seam with strength properties similar to those of the core material is a scientific challenge: a high-density structure presupposes prolonged hold time of the material at elevated temperatures, while a UFG structure may be preserved only if grain growth is limited by means of reducing the heating temperature. This problem may be fixed if the conditions for high-rate heating of UFG materials are extended to include additional mechanisms of diffusion mass transfer related to nonequilibrium grain boundaries.

The goal of the paper is to study the SPS potential for diffusion welding of high-strength corrosion-resistant UFG titanium alloys used in nuclear engineering.

2. Materials. Experimental techniques

Samples of UFG titanium pseudo-α alloy Ti-4.73Al-1.88V (Russian industrial alloy PT3V) served as the materials of research. The UFG structure in materials under study was formed through Equal Channel Angular Pressing (ECAP). The number of ECAP cycles ranged from 1 to 4, while ECAP temperature varied from 450°C to 475°C. The rate of deformation during ECAP was 0.2-0.4 mm/s.
Diffusion welding was performed using a Dr.Sinter model SPS-625. Samples sized 7x7x3.5 mm obtained through spark cutting from UFG metal pieces sized 14×14×100 mm were welded. The level of surface roughness varied as a result of polishing with a diamond paste of various dispersion level (20-28 μm, 10-14 μm, 3-5 μm).

Diffusion welding was performed at the heating rate varying from 10 to 350°C/min, at the welding temperature ranging from 600 to 1140°C and under the pressure applied within the interval from 50 to 100 MPa. Diffusion welding time (at the temperature selected) varied from 0 to 90 minutes. The value and rate of shrinkage were controlled with a dilatometer being a part of a Dr.Sinter model SPS-625. The temperature was measured with a pyrometer Chino IR-AH focused on the side surface of samples. Structure studies were carried out using Jeol JSM-6490 scanning electron microscope and Jeol JEM-2100 transmission electron microscope. Pore size in weld seams was determined with Leica DM IRM metallographic microscope and Jeol JSM-6490 scanning electron microscope.

Microhardness was measured using Duramin Struers-5 microhardness tester with a 2 kg load. Electrochemical studies of samples were carried out in aqueous solution 10%HNO_3+0.2%HF using R-8 potentiostat-galvanostat (Russia) at room temperature. Analysis of Tafel areas of anodic potentiodynamic dependences 'potential – current density' in semi-logarithmic coordinates helped to determine corrosion current density (i_{cor}) which is proportionate to material corrosion rate in the given environment.

3. Experimental results

In the initial state the alloy is characterized by heterogeneous size distribution of grains. The average grain size varies from 10-20 μm to 50-100 μm (Fig. 1a). The average grain size after N=4 ECAP cycles is 0.2-0.5 μm (Fig. 1b).

Energy dispersive analysis reveals two types of grain boundaries (GB) in the structure of a coarse-grained alloy. Type one includes grain boundaries that we have termed 'clean'. These grain boundaries form an absolute majority. Average local concentration of aluminum in such GB is 3.8±0.9 wt.%, while average vanadium concentration is 1.9±0.2 wt.%. These figures differ little from concentrations of these elements in the material volume: average aluminum concentration in GBs is 3.6±0.9 wt.%, while vanadium concentration approximates ~1.4±0.2 wt.%.

The second GB type includes grain boundaries the local concentration of vanadium in which may reach 10 wt.% (Fig.2a). There are also single boundaries the local concentration of vanadium in which reaches 16-18 wt.%, while aluminum concentration is under 1 wt.%. The average value of aluminum concentration in the crystal lattice near such boundaries approximates ~4 wt.%, while vanadium concentration is ~1.6 wt.%.

It shall be emphasized that vanadium-rich GB are found in the areas of the titanium alloy structure where the average grain size substantially exceeds the average size of a fine-grained matrix (Fig. 1a). Grain boundaries in UFG alloys are clean with no excess vanadium concentrations detected (Fig. 2b). Average vanadium concentration on UFG alloy grain boundaries is 1.9±0.3 wt.%, which is close to vanadium concentration in a crystal lattice 1.5±0.3 wt.%.

IOP Conf. Series: Materials Science and Engineering 218 (2017) 012013 doi:10.1088/1757-899X/218/1/012013

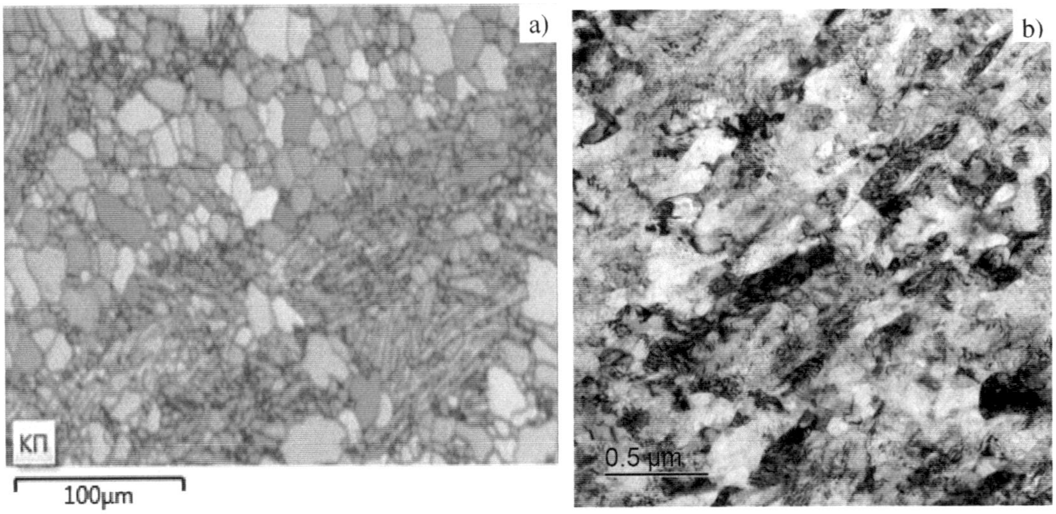

Figure 1. Titanium alloy structure in the original state (a) and after ECAP (b).

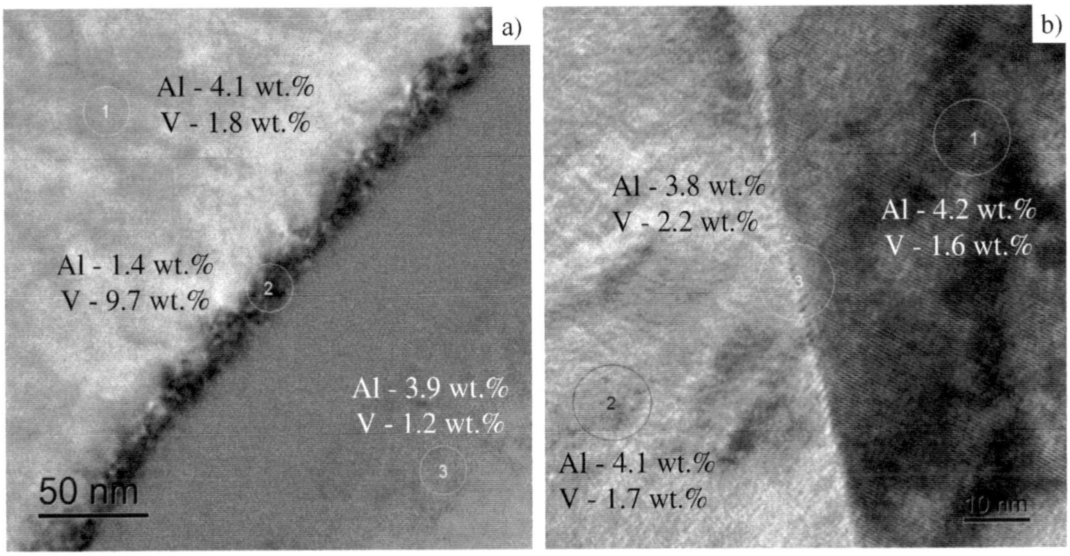

Figure 2. Results of energy dispersive analysis of grain boundaries in a coarse-grained alloy (a) and UFG titanium alloy (b).

Analysis of dependences of the shrinkage value (L) and the shrinkage rate (S) on the heating time prove that the welding process in UFG samples is more intensive: the shrinkage value and shrinkage rate in UFG titanium alloys appear to be much higher than similar characteristics in coarse-grained samples. The shrinkage value in coarse-grained alloys may be smaller than the thermal expansion contribution of the plant and remain unregistered by a dilatometer.

When summarized, the findings prove that under identical temperature and heating rate welding conditions, the average size and volume ratio of pores in UFG samples appear to be much smaller than in coarse-grained materials (Fig. 3).

IOP Conf. Series: Materials Science and Engineering **218** (2017) 012013 doi:10.1088/1757-899X/218/1/012013

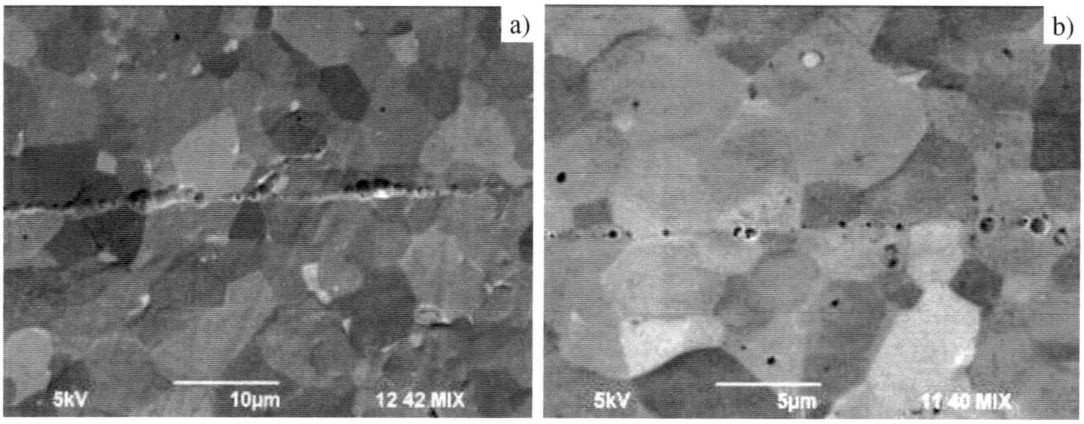

Figure 3. Weld seam produced from coarse-grained alloys (a) and UFG titanium alloys (b) obtained under identical temperature and heating rate welding conditions. Welding temperature 1030°C (above recrystallization temperature). Heating rate V=50°C/min, pressure 50 MPa.

The studies conducted show that the rate of weldability (shrinkage rate) in UFG alloys tends to increase if the heating rate grows. With coarse-grained alloys we observe an opposite effect: an increase in the heating rate leads to a decrease in the shrinkage rate (intensity of the welding process) (Fig. 4). We reckon that this effect arises from the following: high-rate heating (caused by overall reduction in the process time) brings down the intensity of grain boundaries migration and enables a more fine-grained structure (as compared to slower heating). This in turn decreases the typical scale of diffusion mass transfer x, the value of which in case of grain boundary diffusion is proportionate to grain size.

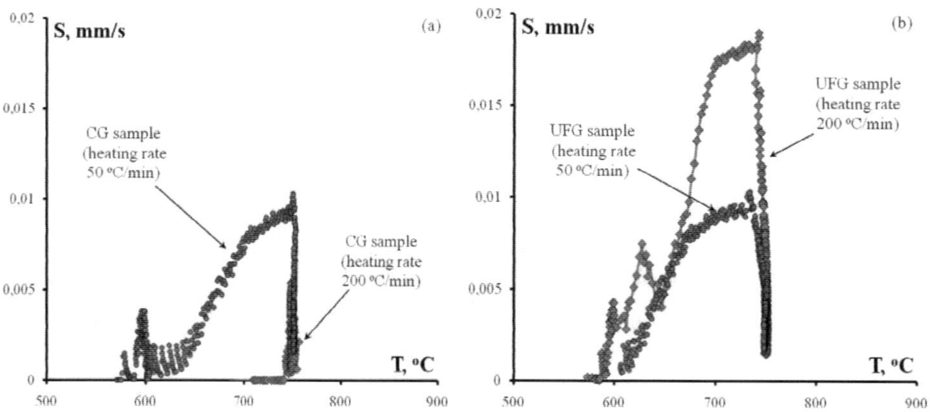

Figure 4. Comparison of temperature dependences on shrinkage rates for coarse-grained (CG) alloy samples (a) and UFG titanium alloy samples (b) with initial roughness level of 3-5 µm welded at heating rates of 50°C/min and 200°C/min under 80 MPa.

The research conducted into mechanical properties proves that the hardness of weld seams obtained though SPS (Fig. 5a) at temperatures below the temperature at the start of recrystallization process corresponds to the hardness of the core material measured at a distance of no less than the thickness of three walls of a welded item from the seam edge (Fig. 5b).

For the purposes of comparison, Fig. 5 (c, d) contains photos showing weld seams in UFG alloy samples obtained through argon arc welding (Fig. 5c) and electron beam welding (Fig. 5d).

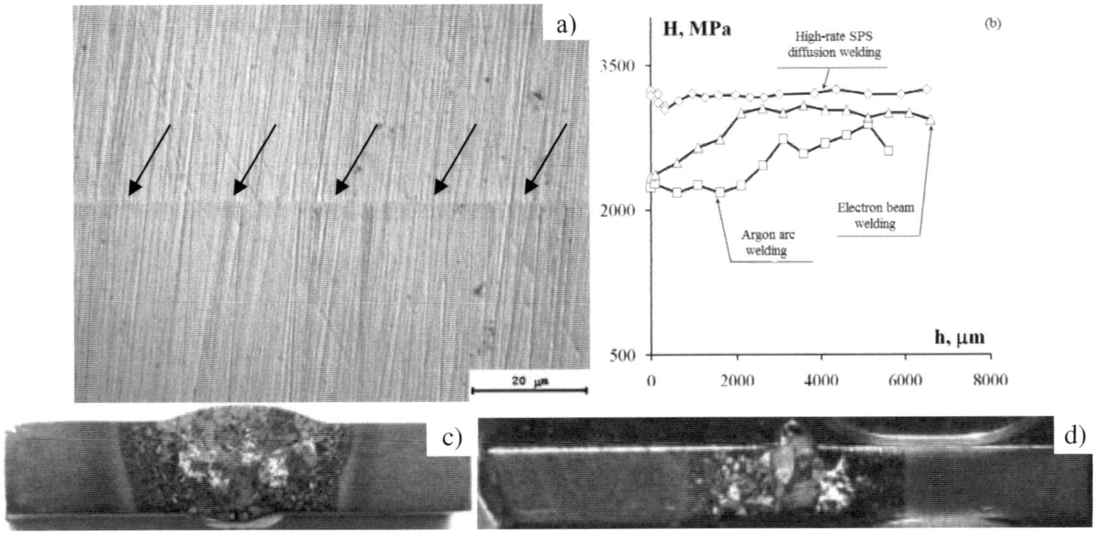

Figure 5. Microsection of a weld seam obtained though SPS (a), argon arc welding (c) and electron beam welding (d) of UFG alloy samples, microhardness distribution (b) in a cross section of UFG alloy weld seams obtained through a variety of welding technologies (h – distance from the seam center).

Table 1. Results of electrochemical studies of weld seams produced from coarse-grained and UFG titanium alloy (heating rate 10°C/min, pressure 50 MPa).

State	SPS welding temperature (°C)	Maximal current density (mA/cm^2)
Initial state (Coarse-grained)	600	1.15
	700	1.70
	800	0.25
UFG state (ECAP, N=4)	600	0.70
	700	1.00
	800	0.12

When comparing anodic potentiodynamic dependences for CG and UFG alloy samples after diffusion welding it should be noted that corrosion current density of weld seams from UFG titanium alloy within the temperature range of 600°C to 800°C appears to be less than current density of CG alloy weld seams ensured through similar welding modes.

With an increase in diffusion welding temperature, the difference in corrosion resistance between weld seams from CG and UFG titanium alloys grows: after diffusion welding at the temperature of 600°C, the ratio of corrosion current density in weld seams produced from CG and UFG alloys was ~1.6, however when the welding temperature rose to 700°C and 800°C this ratio increased to 1.7 and ~2.1 respectively.

In our opinion, this effect is associated with recrystallization processes which underlie formation of a more homogeneous equiaxed structure in UFG titanium alloys, as well as with reduced local concentration of corrosive additives (V, Al) on UFG titanium alloy grain boundaries (see Fig. 2).
Analysis of anodic potentiodynamic dependences for UFG alloy weld seams obtained at different heating rates shows that irrespective of the original roughness, corrosion current density in samples obtained at a high heating rate (200°C/min) appears to be ~1.5 times less than in samples obtained at the heating rate of 50°C. Thus, corrosion current density i_{cor} for UFG alloy weld seams (with initial roughness of 3-5 μm) obtained at the heating rate of 50°C/min is 0.98 mA/cm^2, and at the heating rate of 200°C/min it equals 0.5 mA/cm^2. Current density for CG alloy weld seams obtained under similar conditions appears to be higher: for weld seams obtained at the heating rate of 50°C/min, the i_{cor} ~2.5 mA/cm^2, while those obtained at the heating rate of 200°C/min, the i_{cor} ~2.8 mA/cm^2.

Thus, the corrosion resistance of weld seams in UFG metal samples appears to be higher than the corrosion resistance of weld seams in CG materials.

4. Conclusion

SPS is proven to be an effective technology for high-rate diffusion welding of UFG titanium alloys. Optimization of temperature and rate conditions for SPS allows to produce UFG titanium alloy weld seams the hardness of which corresponds to that of the core material, whereas corrosion resistance is superior to that of the coarse-grained alloy weld seams.

Porosity of weld seams from UFG titanium alloys is found to be less than porosity of weld seams from coarse-grained titanium alloys obtained under similar conditions. An increase in the heating rate leads to higher intensity of diffusion welding with UFG titanium alloy samples, however with coarse-grained alloys higher heating rates result in reduced intensity of diffusion welding.

This effect is associated with the formation of UFG structure in a weld seam during high-rate diffusion welding, which helps to reduce the typical scale of diffusion mass transfer and typical time of diffusion 'exit' of pores from the weld seam due to the contribution of grain boundary diffusion the intensity of which significantly exceeds the diffusion intensity in a crystal lattice limiting the rate of diffusion welding of coarse-grained materials.

Acknowledgements

The authors are grateful for the support to the Russian Science Foundation (grant No. 16-13-00066). The authors thank the Afrikantov Experimental Design Bureau for Mechanical Engineering (Nizhny Novgorod) for conducting argon arc welding and electron beam welding of UFG titanium alloys.

References

[1] Segal V M, Beyerlein I J, Tome C N, Chuvil’deev V N, Kopylov V I 2010 *Fundamentals and Engineering of Severe Plastic Deformation* (New York: Nova Science Publishers) 542 p.
[2] Valiev R Z, Langdon T G 2006 *Progress in Materials Science*. **51** (**7**) 881.
[3] Wang Y, Chen M, Zhou F, Ma E 2002 *Nature* **419 (6910)** 912.
[4] Chuvil'deev V N, Kopylov V I, Gryaznov M Yu, Sysoev A N, Ovsyannikov B V, Flyagin A A 2008 *Doklady Physics* **53 (11)** 584.
[5] Chuvil’deev V N, Kopylov V I, Bakmet’ev A M, Sandler N G, Nokhrin A V, Tryaev P V, Lopatin Yu G, Kozlova N A, Piskunov A V, Melekhin N V 2012 *Doklady Physics* **57 (1)** 10.
[6] Andrievskii R A 2011 *Nanotechnologies in Russia* **6 (5-6)** 357.
[7] Chuvil’deev V N, Shavleva A V, Nokhrin A V, Pirozhnikova O Ed, Gryaznov M Yu, Lopatin Yu G, Sysoev A N, Melekhin N V, Sakharov N V, Kopylov V I, Myshlayev M M 2010 *Physics of the Solid State* **52 (5)** 1098.
[8] Chuvil’deev V N, Nieh T G, Gryaznov M Yu, Sysoev A N, Kopylov V I 2004 *Scripta Materialia* **50 (6)** 861.
[9] Perevezentsev V N, Chuvil'deev V N, Kopylov V I, Sysoev A N, Langdon T G 2002 *Annales de Chimie: Science des Materiaux* **27 (3)** 36.
[10] Tokita M 2013 *Spark Plasma Sintering (SPS) Method, Systems, and Applications (Chapter 11.2.3)*. In Handbook of Advanced Ceramics (Second Ed.). Ed. Shigeyuki Somiya, Academic Press. 1149.
[11] Orlova A I, Volgutov V Yu, Mikhailov D A, Bykov D M, Skuratov V A, Chuvil’deev V N, Nokhrin A V, Boldin M S, Sakharov N V 2014 *Journal of Nuclear Materials* **446** (**1-3**) 232.
[12] Chuvildeev V N, Panov D V, Boldin M S, Nokhrin A V, Blagoveshensky Yu V, Sakharov N V, Shotin S V, Kotkov D N 2015 *Acta Astronautica* **109** 172.
[13] Chuvil’deev V N, Boldin M S, Dyatlova Ya G, Rumyantsev V I, Ordan’yan S S 2015 *Inorganic Materials* **51** (**10**) 1047.
[14] Chuvil’deev V N, Nokhrin A V, Baranov G V, Moskvicheva A V, Boldin M S, Kotkov D N, Sakharov N V, Blagoveshensky Yu V, Shotin S V, Melekhin N V, Belov V Yu 2013 *Nanotechnologies in Russia* **8 (1-2)** 108.

IOP Conf. Series: Materials Science and Engineering **218** (2017) 012014 doi:10.1088/1757-899X/218/1/012014

Structure and characteristics of functional powder composite materials obtained by spark plasma sintering

S A Oglezneva, M N Kachenyuk, V B Kulmeteva and N B Ogleznev

Science Center of Powder Material Science of Perm National Research Polytechnic Univercity, 6 Professora Pozdeeva street, Perm, 614013, Russia

E-mail: director@pm.pstu.ru

Abstract. The article describes the results of spark plasma sintering of ceramic materials based on titanium carbide, titanium carbosilicide, ceramic composite materials based on zirconium oxide, strengthened by carbon nanostructures and composite materials of electrotechnical purpose based on copper with addition of carbon structures and titanium carbosilicide. The research shows that the spark plasma sintering can achieve relative density of the material up to 98%. The effect of sintering temperature on the phase composition, density and porosity of the final product has been studied. It was found that with addition of carbon nanostructures the relative density and hardness decrease, but the fracture strength of ZrO_2 increases up to times 2. The relative erosion resistance of the electrodes made of composite copper-based powder materials, obtained by spark plasma sintering during electroerosion treatment of tool steel exceeds that parameter of pure copper up to times 15.

1. Introduction

Technical ceramics, due to its high hardness, wear resistance and temperature resistance is widely used in modern technology, performing a variety of functions. At the same time, production of high density ceramic materials presents a problem as sintering of the materials requires high temperatures and, in some cases, mechanical pressure, which greatly increases the cost of production. Spark plasma sintering is currently regarded as one of the fastest and most efficient way to consolidate various powder materials, including ceramic materials. In addition to the cost efficiency, spark plasma sintering has another advantage: due to the speed of the process, it can support preserving the structure of nanocrystalline structure of material. [1].

Until now, the factors hindering the expansion of application of ceramic materials has been their higher fragility, low resistance to thermal shock and mechanical unreliability. Currently the ceramic functional materials with special properties are explored worldwide. One of such materials is titanium carbosilicide, which possesses properties of ceramic as well as metal. Like metals it has excellent electrical and thermal conductivity, processability, resistance to thermal shock. The ternary titanium carbosilicide also replicates some ceramic properties: good oxidation resistance up to 1400°C and relatively low density (~4.52 g/cm3). [2, 3].

It is possible to raise the fracture resistance and reliability of standard ceramics (e.g. zirconium oxide) by strengthening them using carbon nanomaterials which possess record-breaking mechanical properties. Pioneering studies on reinforcing ceramics by carbon nanotubes (CNT) have shown an increase of strength and fracture resistance, as well as changes in thermal and electrical conductivity and other properties of such composite materials [4-5].

The use of spark plasma sintering for consolidation of copper-based composite material can reduce the temperature and time of sintering and obtain a product of a given configuration.

2. Experimental procedure

To produce ceramic composite material the powder ZrO_2 was used, stabilized by 3 mol. % of Y_2O_3, (DeguDent GmbH), with the average particle size of 65±10 nm. The synthesis of CNT was carried out at the CVDomna system set (NanoDevice Technology, Russia) at the temperature of 650 °C by the method of by catalytic pyrolysis of ethanol vapor (96 %) with the pressure of 15 kPa to 17 kPa. Matrix powder ZrO_2 was used as the carrier of Ni-Y catalyst used matrix powder. The obtained nanotubes generally are generally multilayer with a diameter less than 20 nm. Purified nanotubes were mixed with powder ZrO_2 in ethanol for 4 hours.

The particles of multilayer graphene were obtained by liquid phase separation of expanded graphite produced by JSC "Silur" (Perm, Russia), which was conducted with an ultrasonic disperser UZDN-2T (400 Wt, 22 kHz) in aqueous solution of sodium dodecyl sulfate. Mixing of ZrO_2-3Y_2O_3 powder with graphene-containing suspension was carried out in aqueous solutions of surfactant for 10 hours. Powder materials of various composition were subjected to consolidation after the mixing (Table 1).

Table 1. Composition and parameters of powder mixtures.

№ composition	Composition	The rotation speed of the mill at MA (min^{-1})	Proportions of the milling body and the charging material	Environment
1	71 wt. % Ti + 25 wt. % SiC + 4 wt % graphite	280	7,5:1	Vacuum
2	90 wt. % TiC + 10 wt. % SiC			
3	80 wt. % TiC + 20 wt. % SiC			
4	60 wt. % TiC + 40 wt. % SiC			
5	ZrO_2-3 mole%Y_2O_3 + (0.5-3) wt. % CNT	160	5:1	C_2H_5OH
6	ZrO_2-3 mole% Y_2O_3+ (0.5-1) vol.% graphene			SDS
7	Cu + (25 – 37.5) vol. % Ti_3SiC_2	160	1:3	C_2H_5OH
8	Cu + (1-5) vol. % colloidal graphite			

Spark plasma sintering was carried out on the Dr. Sinter SPS-1050b system (SPS Syntex, Inc., Japan). The heating was conducted by passing pulsed direct current at the heating rate of 50 °C/min to 80 °C/min. The temperature was monitored along the outside of the graphite matrix with an optical pyrometer. The material was loaded immediately before the heating by constant load (at the pressure equal to 15 MPa and 30 MPa), which was removed after the heating. Isothermal exposure was equal to 1, 5, 10, 25 and 30 min.

The density of compact samples was determined by hydrostatic weighing. The porosity was calculated by the formula $P = (1 - \rho_{exp}/\rho_{theor})\cdot 100$. The theoretical density value was calculated from the phase composition of the sintered material.

X-ray analysis of the synthesized powders and materials based on them was carried out on the diffractometer XRD-6000 (Shimadzu) in λ-Co radiation. Calculation of the phase content was carried out by the PowderCell software. Raman scattering specters were recorded with the multifunctional spectrometer «SENTERRA» (Bruker) at the emission wavelength of 532 nm. Microstructural analysis of fracture surfaces of samples was performed with scanning electron microscopes TESCAN VEGA (Czech Republic) and ULTRA 55 (Carl Zeiss, Germany).

The hardness and fracture resistance of the samples was determined by indentation of polished surface using a Vickers pyramid with a Micro Macro Vickers Hardness tester KBW 10-V (KB Prüftechnik, Germany) at the loading of 98,1 N.

3. Experimental results

Synthesis of compounds of titanium, silicon and carbon occurs during the spark plasma sintering (SPS) of mechanically activated powder composition № 1. The obtained materials is based on titanium carbosilicide Ti_3SiC_2, which quantity depends on temperature and exposure duration during the sintering. The maximum amount of titanium carbosilicide was 86 %. In addition to titanium carbosilicide the sample contained titanium carbide and a minor amount of silicon carbide.

The density of material increases with the sintering temperature increase to 1300 ° C, and after that almost doesn't change. Increasing the exposure duration also causes a small increase of density and decreases the porosity. The minimal porosity of the material based on titanium carbosilicide was 2%.

The structure of the material consists of two main structural element: titanium carbosilicide grain of lamellar structure the size of 1 µm to 10 µm and titanium carbide grain the size of 0,5 µm to 10 µm. The microstructure is non-homogeneous and contains areas of even distribution of titanium carbosilicide and titanium carbide, as well as with predominant content of one phase (Fig. 1).

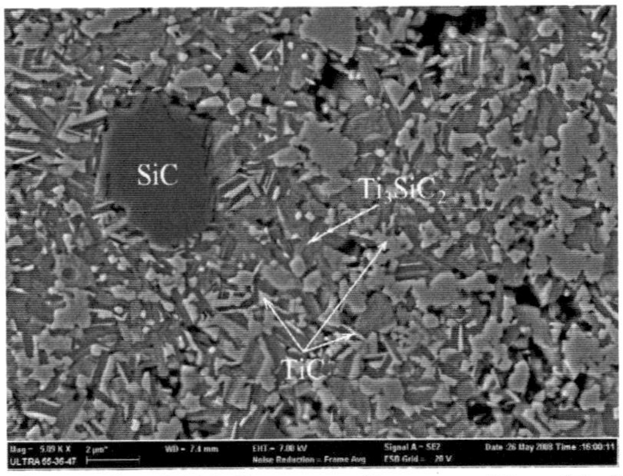

Figure 1. SEM Images of composition № 1 after SPS at 1400° C.

Figure 2. SEM Images of Fractures of TiC + 20 wt.% SiC with magnifications of 2000×

SPS of the compositions № 2, 3 and 4 does not lead to the formation of new compounds, the composition of sintered materials doesn't differ from the composition of the initial charge. With increase of temperature of SPS the porosity of the compound TiC - 10% SiC decreases almost linearly. For the other compounds in the temperature range of 1100 ° C to 1200 ° C, the porosity remains almost unchanged, and then decreases. The compound TiC - 10% SiC sintered at the temperature of 1400 °C has the least porosity equal to 4%.

The grains of the matrix (titanium carbide) are substantially smaller than the grains of silicon carbide (ones and tens of micrometers, respectively) (Figure 2). Separate grains tightly adjoin one another, which explain the residual porosity of the material, the size of most of the pores is less than one micrometer.

The intensity of wearing of the material № 3, obtained at the SPS temperature of 1350 °C, during the testing by the "disk-block" scheme with dry friction at the pressure of 1.5 MPa equaled $3.8 \cdot 10^{-9}$ mm wear/mm rubbing path, which is five times lower than that of silicon carbide obtained by hot pressing.

Spark plasma sintering of composite materials based on ZrO_2, strengthened by CNT and nanoplates of multilayer grapheme, was carried out at the temperature of 1300 °C with isothermal exposure of 30 min. According to data of Raman spectroscopy the matrix of the composite materials consist of tetragonal ZrO_2. It was established that with an increase in content of CNT composite densifying processes are slowing, the shrinkage is reducing, ending for all samples at temperatures about 1100 °C. The same regularity is observed for the material with addition of multilayer graphene, but it reduces the shrinkage during sintering to a slightly greater extent than CNT. Table 2 presents the results of measurement of hardness and fracture resistance of the obtained samples.

Table 2. Properties of ceramic material ZrO_2-$3Y_2O_3$ with carbon nanostructures after the sintering.

The content and form of the carbon nanostructures	Sintering mode			*P* (%)	HV (GPa)	K_{1C} (MPa·m$^{1/2}$)
	T (°C)	t (min)	P (MPa)			
0	1350	120	0 (no SPS)	0	13,7±0,1	4,9±0,2
0.5 CNT, wt.%	1300	30	30	0,9	12,4±0,1	10,7±0,5
1.0 CNT, wt.%				0,8	10,4±0,1	7,6±0,4
3.0 CNT, wt.%				4,0	9,6±0,1	6,0±0,5
0.5 vol.% graphene				4	12,0±0,2	5,3±0,2
1 vol.% graphene				7	7,2±0,6	7,6±0,8

With the introduction of the CNT the porosity of the composite material increases and the hardness decreases, and the fracture resistance increases, especially with the CNT content equal 0.5 wt.%. The conducted electron microscopic analysis of the fracture surfaces of the sintered samples showed that the nanotubes distribution in the volume of the matrix is not uniform. In addition to CNT located in the grain boundaries, there are also areas containing agglomerates of nanotubes, the amount of which increases with an increase of their content. Apparently, the used parameters of mixing are insufficient for a homogeneous distribution of nanotubes in the matrix, which influences the properties of the composite (Figure 3, a). Introduction of graphene material also reduces the hardness of the composite ceramic matrix, particularly when the content amounts to 1 vol.%. After sintering the hardness of the composite with 0.5 vol.% graphene approaches that of pure zirconium oxide, but with an increase of content of graphene to 1 vol.% it reduced by almost a half. However, increasing the graphene content leads to increased fracture resistance. If with graphene content of 0.5 vol.% the fracture resistance of the composite and the zirconium oxide are substantially equal, the composite with 1 vol.% of graphene has the fracture resistance which is 1.5 times higher.

The graphene particles in form of plates are mostly uniformly distributed in the volume of the ceramic matrix. Among the graphene particles of uniform thickness there are quite large particles with a porous structure like a crumpled paper. Apparently, these are not completely exfoliated particles of initial expanded graphite. Such particles can serve as additional stress concentrators and reduce the positive effect on strengthening exerted by introducing graphene particles into the ceramic matrix. It should be noted that when the graphene particles consist of a quantity of layers, they are arranged in a ceramic matrix as straight plates, and single-layer graphene tends to curtail (Fig. 3, b).

a b

Figure 3. SEM images of fractures of ZrO_2-$3Y_2O_3$: (a) – 1 wt.% CNT; (b) – 1 vol.% graphene

After the spark plasma sintering of “copper-colloidal graphite” material at 900 ° C for 5 min., the signs of formation of diamond links were detected in the graphite, consisting in the appearance of additional reflexes on the diffractogram of graphite with odd *l* (001) and (003) and in splitting of some existing reflexes (002), (004), (006); results are consistent with the data of authors of [6]. In [6] by intercalation of graphite with copper, which after the release of copper ions from the interlayer spaces of the graphite results in a corrugation of graphene planes with formation of sp^3 bonds. Raman spectroscopy of colloidal graphite sintered with copper confirms the formation of sp^3 bonds: a weak D4 peak appears, which indicates the presence of sp^3 bonds, the peaks *G* and *D1* become broader, reflecting the emergence of disordered graphite, and at the *G* peak appears a "shoulder» D2 - so-called "the shoulder of defects", which together with D1 indicates the defects appearing at the edges and inside the graphene planes (Table 3).

Table 3. Raman specters of colloidal graphite and composite material after SPS.

Material	Raman shift (sm^{-1}) (index –Raman absol. intensivity)						
	G	*2D*	*D1*	*D2*	*D4*	*D3*	*LA*
graphite	1580_6	2682_1	1352_4	no	no	1510_1	no
Cu -20 vol. graphite (SPS)	1576_5	no	1351_5	1620_2	1230_1	1510_1	no

Energy-dispersive and X-ray diffraction analysis showed that after the sintering of the "copper-Ti_3SiC_2” material occurs a conversion of about half of the grain titanium carbosilicide to solid solution of carbon based on titanium silicide Ti_5Si_3(C) and small amounts of titanium carbide, silicon carbide, titanium silicide $TiSi_2$ (Figure 4)

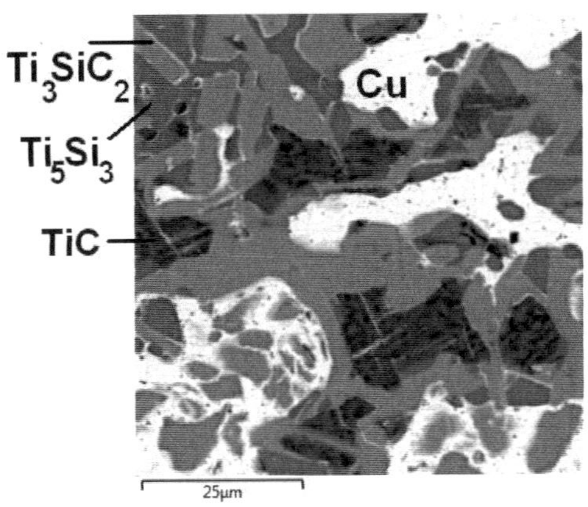

Figure 4. The microstructure of the "copper-Ti_3SiC_2" material

The content of silicon in the grains of titanium carbosilicide is below stoichiometric value by about 14% due to the silicon deintercalation. The electrical resistance of the composite material increased with increasing of concentrations of titanium carbosilicide due to formation of solid solutions in the copper of titanium silicide with low conductivity (Table 4). The performance properties of composite materials after the SPS were better than those of pure copper due to dispersion strengthening and retaining the molten copper in the dispersed pores of carbosilicide.

Table 4. The properties of the sintered powder material "copper-Ti_3SiC_2".

The concentration addition (vol. %)	Porosity (%)	Hardness, HB (MPa)	Flexural strength, (MPa)	Electrical resistivity resistance, (Ohm· m)	Relative erosional wear at the power, %	
					75 mJ	100 mJ
0	0	400	500	0,021	5,6	6,9
0,4 graphite	0	-	-	0,015	0,8	3,5
4,0 graphite	3	-	-	0,023	0,6	3,2
12,5 of Ti_3SiC_2	2	650	280	0,06	3,2	5,2
37,5 of Ti_3SiC_2	7	800	340	0,14	1,0	4,0

4. Conclusion

Ceramic and metallic composite materials of various purposes were obtained by the method of spark plasma sintering; the patterns of densification, phase and structure formation were defined. It has been established that SPS allows obtaining carbon-based ceramic materials with increased wear resistance. Adding carbon nanostructures to zirconium dioxide slightly decreases the density, but increases the fracture resistance by 1.5-2 times. Electrotechnical materials based on copper with additions of titanium carbosilicide or graphite possess significantly better erosion resistance compared to copper.

The research was financially supported by the Ministry of Education and Science of Russian Federation, research project №14.574.21.0065 (RFMEFI57414X0065) – “Research and development of technologies for production and processing of functional nanostructured wear-resistant material on the basis of titanium ternary carbide for engineering products”. The results are shown in Part 4 of the report.

References

[1] Tokita M 2013 *Handbook of Adv. Ceram. (Second Edition): Materials, Applications, Processing, and Properties* Spark Plasma Sintering (SPS) Method, Systems, and Applications **1** 1149-77

[2] Barsoum M W 2000 *Prog. Solid State Chem* The $M_{N+1}AX_N$: A new class of solids Thermodynamically stable nanolaminates **28** 201–281

[3] Barsoum M W and El-Raghy T 1996 J. Am. Ceram. Soc. Synthesis and characterization of a remarkable ceramic: Ti_3SiC_2 **7** 1953–56

[4] Ahmad I, Yazdani B and Zhu Y. 2015 *Nanomaterials* Recent Advances on Carbon Nanotubes and Graphene Reinforced Ceramics Nanocomposites **5** 90-114

[5] Datye A, Wu K-H, Gomes G, Monroy V, Lin H-T, Vleugels J and Vanmeensel K 2010 *Composites Science and Technology* Synthesis, microstructure and mechanical properties of Yttria Stabilized Zirconia (3YTZP) – Multi-Walled Nanotube (MWNTs) nanocomposite by direct in-situ growth of MWNTs on Zirconia particles **14** 2086–92

[6] Andreeva V D, Stepanova T R 2002 *Technical Physics Letters* Effect of copper atoms in the graphite structure **28**(18) 18-23

IOP Conf. Series: Materials Science and Engineering **218** (2017) 012015 doi:10.1088/1757-899X/218/1/012015

Consolidation of Transparent ALON by Spark Plasma Sintering Methods

N A Rubinkovskiy[1], A G Zholnin[1], E G Grigoryev[1], I V Spirin[1], P L Dobrokhotov[1], M G Isaenkova[1], R S Hafizov[2], V I Yuhvind[3], A G Astashov[4] and I O Pahilo-Daryal[4]

[1]National Research Nuclear University MEPhI, Kashirskoe shosse, 31, Moscow, 115409, Russia
[2] Moscow State University of Geodesy and Cartography, Kazakova st. 15, Moscow, 105064, Russia
[3]Institute of Structural Macrokinetics and Materials Science, Russian Academy of Sciences, Akademika Osipyana st. 8, Chernogolovka, Moscow region, 142432,Russia
[4]A. Baikov Institute of Metallurgy and Materials Science, Russian Academy of Sciences. Leninskii Ave. 49, 119991, Russia

E-mail: fissium@yandex.ru

Abstract. The paper studies the prospects of spark plasma sintering of aluminum oxynitride, determines the optimum modes and parameters of compacting. Optimum modes of preparing a powder mixture for sintering are determined. The structure and properties of sintered samples are studied. The factors affecting the transparency of the finished products are ascertained, and the ways that allow obtaining a high light transmittance are identified.

1. Introduction

There is a currently observed rise of penetrative power of small arms and small caliber artillery. That is forcing ammunition developers to seek new opportunities to improve performance of armor materials and protective structures. It especially concerns transparent armor materials. One of the most promising material in this class is transparent polycrystalline ceramics, such as ceramics based on aluminum oxynitride (ALON). This material possesses a unique combination of properties required for use in armored protection devices and furthermore does not require the use of overly expensive technologies for production. Manufacturing transparent product of complex shape from this material is possible by using of molding and sintering traditional for ceramic technology processes. Development of armored protection made of the aluminum oxynitride attracts an active international interest, especially from the companies working for defense industries.

Among all the ceramics nowadays an average density aluminum oxynitride has a significantly high strength comparable to the YAG (yttrium aluminum garnet) and cubic zirconia (stabilized zirconium oxide). It is also the most advanced armored protection comparing its impact strength. ALON surpasses all transparent materials, including quartz glass (fused quartz, spinel and leucosapphire).

IOP Conf. Series: Materials Science and Engineering **218** (2017) 012015 doi:10.1088/1757-899X/218/1/012015

Aluminum oxynitride is a compound of aluminum, oxygen and nitrogen described by the formula $Al_{23}O_{27}N_5$. This compound was synthesized as a powder quite recently - in the 50s. In the late 80s and early 90s a technology for producing transparent products of complex shape was developed. In Russia no attempts of such production have been made yet. Nowadays the interest for the aluminum oxynitride has greatly increased. With low crystal symmetry – cubic crystal system - the material makes it possible to produce transparent ceramics using conventional ceramic technology.

Aluminum oxynitride is usually made from powder compounds of aluminum oxide and aluminum nitride. There are two main objectives in the sintering process: to provide a single-phase material, and to minimize porosity. Pores and second phases are the most intense centers of light scattering. These requirements does not apply to conventional ceramic materials.

The main objectives of this research were choosing the raw material, determination of the optimum sintering methods and modes for aluminum oxynitride, and identification of factors affecting its transparency.

2. Research methods and equipment

Micrographs of the surface structure of the obtained samples were examined with an optical metallographic microscope MMP-3. For obtaining a homogeneous mixture of the oxide and nitride powders the ultrasonic dispersing in alcohol with an ultrasonic disperser Sonicator Q500 on 50% of its capacity was used. Duration of dispersion was 40 min.

Spark plasma sintering (SPS) was conducted in a LABOX machine model 625 (Sinter Land, Japan). The sintering procedure was standard: the graphite matrix with an additional layer of graphite paper between the matrix and the powder filling, and between the graphite punches and the powder to prevent the latter and the equipment from sintering. The temperature control was performed with an optical pyrometer with a measuring range 573°C to 3000 °C, directed to the matrix through a hole cut in the graphite felt, which was used as thermal insulation to prevent heat loss by radiation.

The sintered pellets were subjected to grinding and polishing by diamond pastes on the automatic polishing machine Mecatech 234. The microhardness of the samples was measured by the Vickers hardness test with a microhardness tester FM-800 (Future-Tech). The density of the samples was calculated according to the data of hydrostatic weighing with a high-accuracy analytical balances 210 G X 0.1 Mg

The bending strength characteristics of the sintered disk sample were measured on a universal testing machine 50 Quasar. A disk sample of height *h*, placed on a support ring, was forced by the truncated cone-shaped punch. Maximum load P was assumed for calculating the ultimate strength prior to the sample failure by the formula [1].

$$\sigma = \frac{3P}{8\pi h^2}\left[4-(1-\mu)\left(\frac{d}{D}\right)^2+4(1+\mu)\ln\left(\frac{D}{d}\right)\right] \quad (1)$$

where d – punch diameter, mm, μ - Poisson's ratio, *D* – diameter of the bearing ring.

X-ray phase analysis was conducted on the D8 Discover system. The break structure of the samples was studied with a scanning electron microscope (SEM) TESCAN (model VEGA 3).

3. Experimental results

3.1 Spark plasma sintering of aluminum oxynitride powder

In the first stage the sintering was conducted on already prepared powders obtained in the mode of self-propagating high-temperature synthesis (SHS), followed by pulverizing, which were produced by the Institute of Structural Macrokinetics and Materials Science RAS (Russian Academy of Sciences) in Chernogolovka city, Moscow region. The average particle size amounted to 80 microns (Fig. 1).

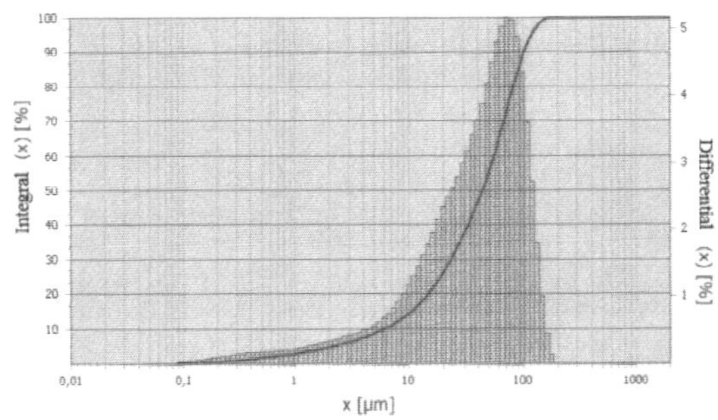

Figure 1. The particle size distribution of the aluminum oxynitride powder produced by the Institute of Structural Macrokinetics and Materials Science RAS

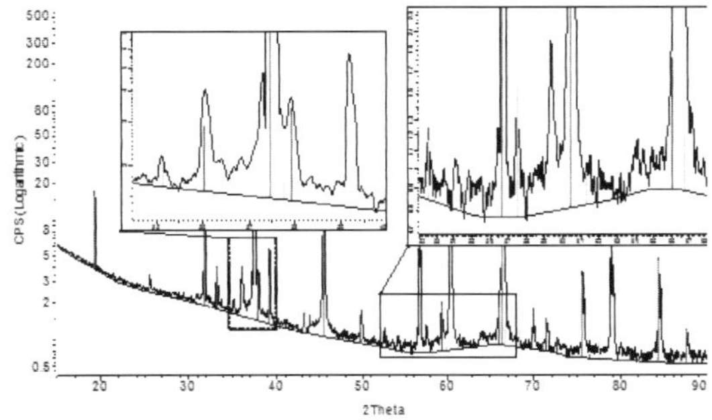

Figure 2. The X-ray spectrum of the aluminum oxynitride powder produced by the Institute of Structural Macrokinetics and Materials Science RAS (on a logarithmic scale)

Figure 2 shows the data of the X-ray analysis, which allows assessing the quality of the supplied aluminum oxynitride powder. According to the results of quantitative analysis, in the aluminum oxynitride powder produced by the Institute of Structural Macrokinetics and Materials Science RAS ISMAN there is a proportion of following phases: ALON - 96% (red), Al2O3 - 1,4% (lime), AlN - 2,6% (blue).

The mode of spark-plasma sintering was selected empirically based on maximum achievable density, microhardness and the presence of transparency. The maximum density (3.64 g / cm3, or 99.1% of the theoretical), and microhardness (18.6 GPa) were obtained with a 10-minute exposure at 1850 ° C, a pressure of 40 MPa, a heating and cooling rate of 100°C/min and 50°C/min respectively. The grain size at 10 minute exposure was equal 95 microns, and with increasing exposure time to 60 minutes the grain size increased to 120 microns, however the density and microhardness decreased. Bending strength in these samples ranged from 214 - 261 MPa, and conditional modulus varied from 172 GPa to 198 GPa. The maximum shrinkage during the sintering occurred at the temperature range 1800°C to 1850 °C. At the same time some transparency was observed in compacts with a diameter of 15 mm sintered in this mode, as illustrated by Figures 3 and 4.

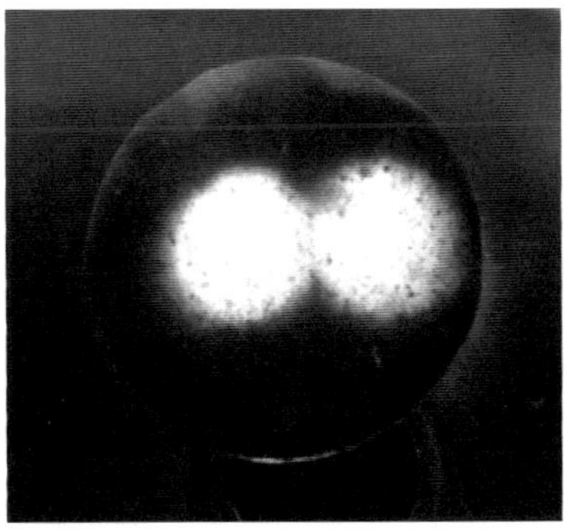

Figure 3. A polished sample, 1 mm thick, positioned on two LEDs

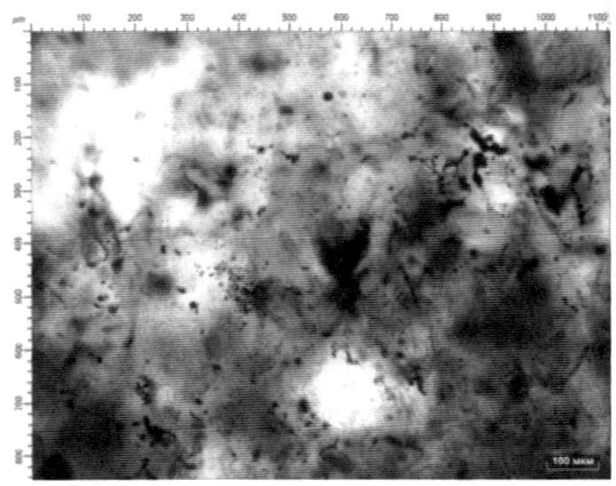

Figure 4. Image of a sample, 1 mm thick from a transmission electron microscope

As seen from the presented figures, there are small nontransparent inclusions along with transparent grains. These nontransparent inclusions may be pores or inclusions of second phase. To determine their nature an X-ray spectrum of the compact was taken. The proportion of phases: ALON - 95,4%, Al7O3N5 - 4,2%, AlN - 0,4%, which is also close to the composition of the starting powder.

The microanalysis of the sample fracture performed on a SEM, which is presented in Figure 5, showed that in contrast to large grains of ALON small inclusions contain admixtures of sulfur and iron, as indicated in Table 1, obviously residual products of thermal reaction preceding the production of ALON.

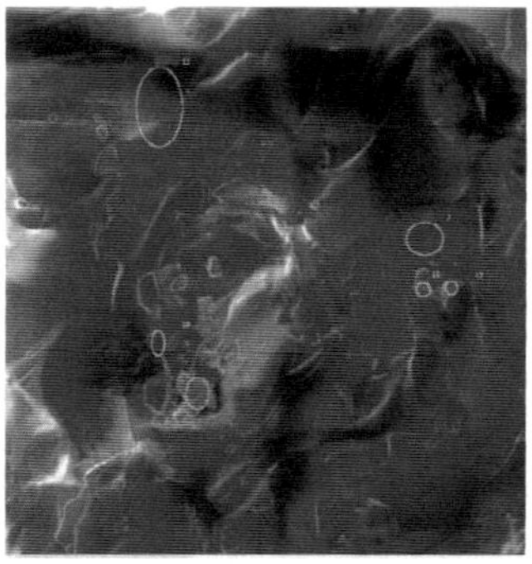

Figure 5. Areas of fracture sample, the chemical composition in which was studied.

Table 1. Chemical composition of studied areas.

№ specter	Chemical composition (at. %)				
	N	O	Al	S	Fe
7	15.64	64.71	19.55	0.09	
9	10.00	52.24	29.00	8.76	
10	12.89	60.60	26.48	0.03	
11	3.16	23.10	21.79	28.15	23.81
12	9.81	63.41	25.59	1.19	

Consequently, the ALON powder produced by the Institute of Structural Macrokinetics and Materials Science RAS, Chernogolovka city, is well sintered, shows acceptable mechanical properties, but doesn't have a sufficient purity to produce a transparent material.

3.2. Solid-phase synthesis of aluminum oxynitride

The phase diagram of Al2O3-AlN system shows that aluminum oxynitride has a sufficiently broad homogeneity region that allows varying the composition of the starting mixture. [2] However, it can be assumed that the material will have the maximum properties in the middle of the range of existence of the γ-aluminum oxynitride phase. For this reason, the starting components were taken by the following proportion: 64,3 at% of Al_2O_3 64,3 at% and 35,7 at.%. of AlN. The powders were mixed by specified proportion in alcohol in an ultrasonic disperser in alcohol for 40 minutes, and then air dried.

Sintering of all the samples was carried out in a nitrogen atmosphere at pressure equal 105 Pa. This is because the nitrogen in aluminum nitride may transit into the gas phase consequently reducing the quantity of AlN. Conversely, because of the ability of aluminum to alter its valence to 1 at high temperatures, there is a possibility that nitrogen will displace oxygen in aluminum oxide. In this case, the divergence of stoichiometry will occur in the opposite direction.

Analysis of the Al2O3-AlN system state chart revealed that the formation of γ-aluminum oxynitride phase starts at 1650 ° C. The literature also shows that 1650 ° C is the temperature at which starting materials completely transform to aluminum oxynitride [3-7]. These were the reasons why the holding temperature was 1700 °C which is slightly above the theoretical value. The available evidence seems to suggest that there is a temperature gradient in diameter of the mold, i.e. areas directly adjacent to the inner walls of the matrix may not undergo the transformation. In addition, a grain has to be grown to size of about 100 microns to ensure transparency. Higher temperature will also contribute to this. Heating and cooling rate remained the same.

IOP Conf. Series: Materials Science and Engineering **218** (2017) 012015 doi:10.1088/1757-899X/218/1/012015

Empirically determined that the optimum pressure during sintering is 40 MPa. Pressure rise does not improve characteristics of compacts. The optimal exposure time at 1700 °C is 30 minutes. The increasing of the exposure time has no significant effect on the properties of materials.

Characteristics of aluminum oxide and aluminum nitride powders synthesized in the Institute of Structural Macrokinetics and Materials Science RAS, by the plasma-chemical method, are shown in Figure 6. Table 2 presents properties of the sintered samples.

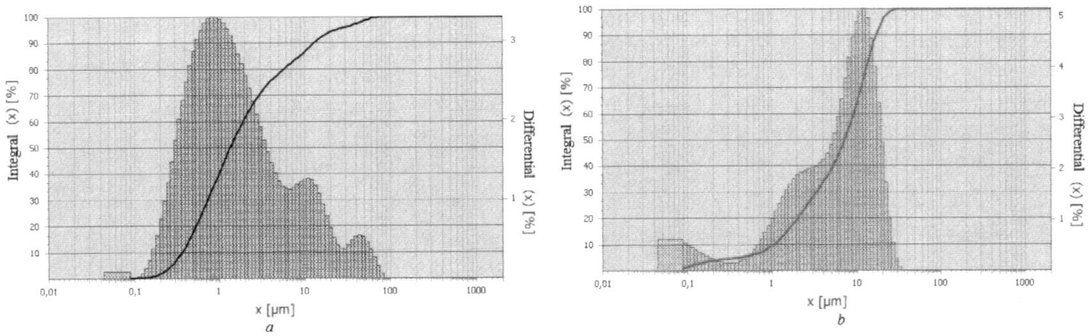

Figure 6. Particle size distribution of the powder: (a) aluminum oxide, (b) aluminum nitride

Table 2. Properties of the sintered samples.

Name	Density (g/cm^3)	The proportion of the theoretical density (%)	The average microhardness (GPa)	Tensile strength (MPa)	The presence of transparency
ALON-1	3.66±0.01	99.8	18.7±0.2	175	+
ALON-2	3.63±0.02	98.9	18.6±0.1	178	+
ALON-3	3.62±0.01	98.1	18.6±0.1	174	+
ALON-4	3.65±0.01	99.2	18.7±0.1	169	+
ALON-5	3.68±0.02	99.9	18.8±0.1	185	+
ALON-6	3.68±0.01	99.9	18.8±0.1	187	+

Samples have transparency, as shown in Figure 7.

Figure 7. Sample ALON-6.

According to the results of the quantitative analysis, the sample contains traces of the starting powders.

4. Discussion of the results

Direct research of the aluminum oxynitride powder produced by the Institute of Structural Macrokinetics and Materials Science RAS) revealed that it contains only 95% of aluminum oxynitride. The same amount of aluminum oxynitride is contained in the sintered samples, i.e. the material is not single-phase. Secondary phases of non-cubic structure act as additional light scattering centers, which leads to a significant reduction in light transmission.

Furthermore, studies with the use of SEM prove the presence of sulfur and iron admixtures in significant quantities, which leads to the formation of additional phases in the material.

Consequently, satisfactory transparency was not achieved due to a high content of secondary phases (according to literature data, small amounts of secondary phases - about 1% - do not introduce significant changes to transparency).

The samples sintered using the solid-phase synthesis reaction, exhibit satisfactory transparency despite the fact that their structure contains secondary phases. This may occur due to a non-optimal ratio of starting powders or the fact that the synthesis reaction has not been fully completed. In addition, carbon can penetrate the material during the sintering. However, the analysis did not confirm its presence, which may be due to the lack of it, as well as the methods applied.

5. Conclusions

The data gathered in this study seem to prove that the spark-plasma sintering of aluminum oxynitride powder obtained by the SHS method on existing technology did not allow obtaining transparent material. The solid-phase synthesis of ALON with the spark-plasma sintering yielded positive results. The use of which makes it possible to produce transparent aluminum oxynitride for the first time in. However, obtaining single-phase transparent material requires further adjustment of the composition of the starting mixture.

Acknowledgment

The work was carried out with the financial support of the Russian Science Foundation (grant № 16-19-10213).

The Interdepartmental laboratory of perspective technologies of creation of new materials at the National Research Nuclear University MEPhI was established under the program of mega-grants of the Russian Government (grant № 11.G34.31.0051).

References

[1] Zholnin A G, Melekhov A P, Hafizov R S, Vakulenko S A, Rubinkovskiy N A, Samokhin A V, Alekseev N V and Grigoryev E G. 2016 Influence of technology of nanopowder production on the microstructure of the sintered by spark-plasma material on the example of aluminum oxide *IOP Conference Series: Materials Science and Engineering*. **130 (1)**

[2] McCauley J, Corbin M High temperature reactions and microstructures in the Al_2O_3 – AlN system 1983 *Progress in Nitrogen Ceramics* The Netherlands. 407

[3] Sahin F C, Kanbur H E, Apak B 2012 Preparation of ALON ceramics via reactive spark plasma sintering *Journal of European Ceramic Society* **32 (4)** 925-929

[4] Li X, Luo J, Zhou Y 2015 Spark plasma sintering behavior of AlON ceramics doped with different concentrations of Y_2O_3 *J. Ceramics International* **35** 2027-2032

[5] Wei W, Fu Z Y, Wang H, Wang W M 2009 Research on AlON transparent ceramics by SPS, *J. Ceramics International* **31** 13-14

[6] Xibao L., Junming L., Ya Z 2015 Spark plasma sintering behavior of AlON ceramics doped with different concentrations of Y_2O_3, *Journal of the European Ceramic Society* **35** 2027–2032

[7] Kaplan S S, Kurama S, Gunkaya S 2015 Spinel nitrides transparent ceramics *Journal of the European Ceramic Society* **35** 3255-3262

IOP Conf. Series: Materials Science and Engineering **218** (2017) 012016 doi:10.1088/1757-899X/218/1/012016

Calculation of thermal fields in a graphite mold during spark-plasma sintering of non-conductive materials

N A Rubinkovskiy, A G Zholnin and E G Grigoryev

National Research Nuclear University MEPhI, Kashirskoe shosse, 31, Moscow, 115409, Russia

E-mail: fissium@yandex.ru

Abstract. The research provides modeling of spark-plasma sintering process of aluminum oxide, specifically the distribution of temperature fields over the sample volume and mold at various stages of heating. Calculation was based on the experimental data on the measurement of temperature on the surface of the mold, inside the mold cavity and in its various internal parts including punches (diameter of 15 mm), in the absence of insulation felt. Experiments established that the main source of heat emission at temperatures less than 1300°C to 1400 °C is contact resistance at the borders of the punches and the matrix molds. Then heat emission zone moves towards the punches. Based on the received data, the temperature and electric parameters were selected for the materials used for the press mold, providing a good agreement between the observed and calculated distribution patterns at different temperatures. Obtained parameters were used in calculations of temperature fields in the press molds with insulation felt and for the press molds with diameter of 15 mm and 10 mm for the production of tablets as well as glazing beads.

1. Introduction

Spark-plasma sintering (SPS) is one of the promising methods for the consolidation of ceramic materials. SPS is based on the combined effect of pulsed direct current and mechanical pressure on the particulate material. The material in the target area is heated to very high temperatures. SPS of the powder material is conducted in a current-conductive graphite press mold, therefore the technology has no restrictions on the type of sintered powder materials. During sintering of conductive powder the heating of the powders and compacts made of them is carried out by passing a direct current through them, and during sintering non-conductive materials - by heating the graphite press mold elements.

For all advantages of the SPS method there is a serious flaw: in the units when the sample temperature is monitored by measuring the surface temperature of the mold or in an opening made in the mold, it is not possible to reliably establish the temperature of the sample itself. This may lead to an underheating, overheating or uneven heating of the sintered material, which in turn will affect the final properties of the compact.

Foreign companies are actively working on the modeling of temperature fields in the mold during spark-plasma sintering, however, most of these research use metallic systems as a sintered material, and are limited to using ceramic materials at low temperatures only [1-6]. No Russian research was found on this issue.

The main objective of this research was the modeling of spark-plasma sintering of aluminum oxide, in particular the distribution of temperature fields over the volumes of the sample and the mold at

IOP Conf. Series: Materials Science and Engineering **218** (2017) 012016 doi:10.1088/1757-899X/218/1/012016

different stages of heating, in order to take into account the possible errors in the production of final products, i.e temperature measurement for various materials with different geometry of compacts.

2. Equipment and methods of research

Spark-plasma sintering (SPS) was conducted in a LABOX machine model 625 (Sinter Land, Japan). The sintering was performed on the standard procedure in the graphite matrix with an additional layer of graphite paper between the matrix and the powder filling, and between the graphite punches and the powder to prevent sintering of the powder to the equipment material. Temperature control was performed with an optical pyrometer with a measuring range 573°C to 3000 °C. directed to the matrix through a hole cut in the graphite felt, which was used as thermal insulation to prevent heat loss by radiation. A previously sintered tablet of 15 mm diameter was used in the measurements. For observation of the temperature distribution inside the mold a quarter of the matrix was removed. For temperature measurements at different surfaces of the system the heating was carried out without using thermal insulation felt. Pictures of the mold were made with a digital camera. Modeling was carried out with a COMSOL Multiphysics software package.

3. The reference experiments

Figure 1 shows the scheme of structures used in the experimental temperature measurements.

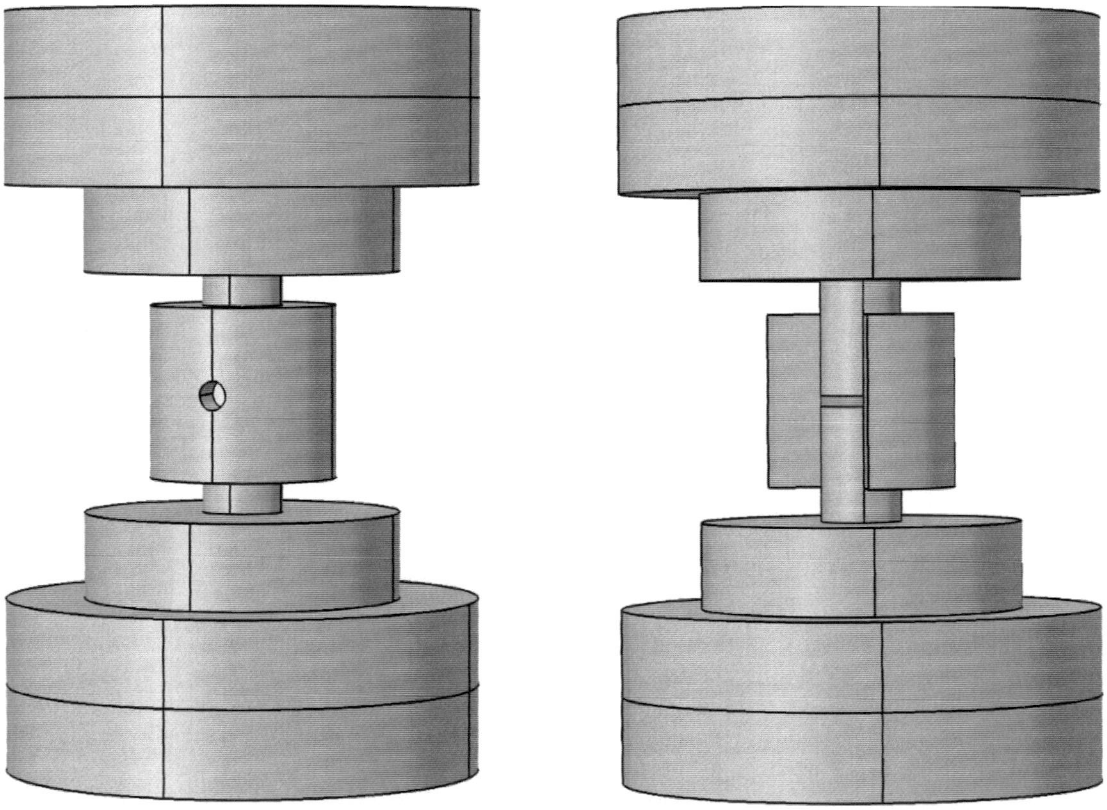

a *b*

Figure 1. Scheme of structures for temperature measurement: (a) an opening diameter of 5 mm and depth of 8 mm, (b) a quarter of matrix is removed.

IOP Conf. Series: Materials Science and Engineering **218** (2017) 012016 doi:10.1088/1757-899X/218/1/012016

Figure 2 shows a photograph of the structure (Figure 1 (a)) at surface temperature equal 1000 ° C and graph of dependence of the difference between the temperature of the opening and the surface temperature on the surface temperature of the press mold.

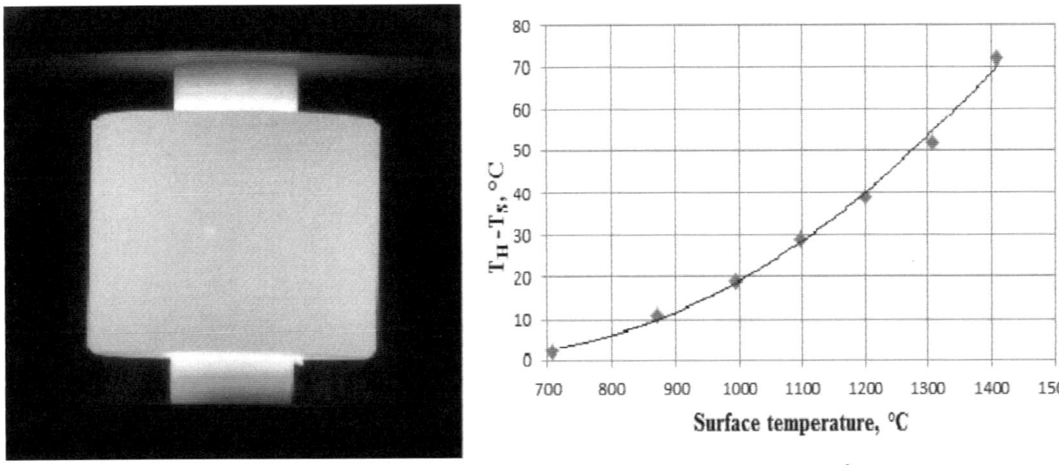

a b

Figure 2. Photograph of the press mold (a); graph of dependence of the difference between the temperature of the opening and the temperature of the surface on the temperature of the surface (b).

The figure 2 shows that as the temperature of the surface increases the temperature difference between the opening and the surface, as expected.

The next step was to determine the difference of temperatures between the surface and the sample. To do that, the experiments were conducted on the structure in Figure 1 (b). Figure 3 shows photographs of the press mold at high and low temperature, and Figure 4 presents a graph of the dependence of the difference between the sample temperature and the surface temperature on the surface temperature.

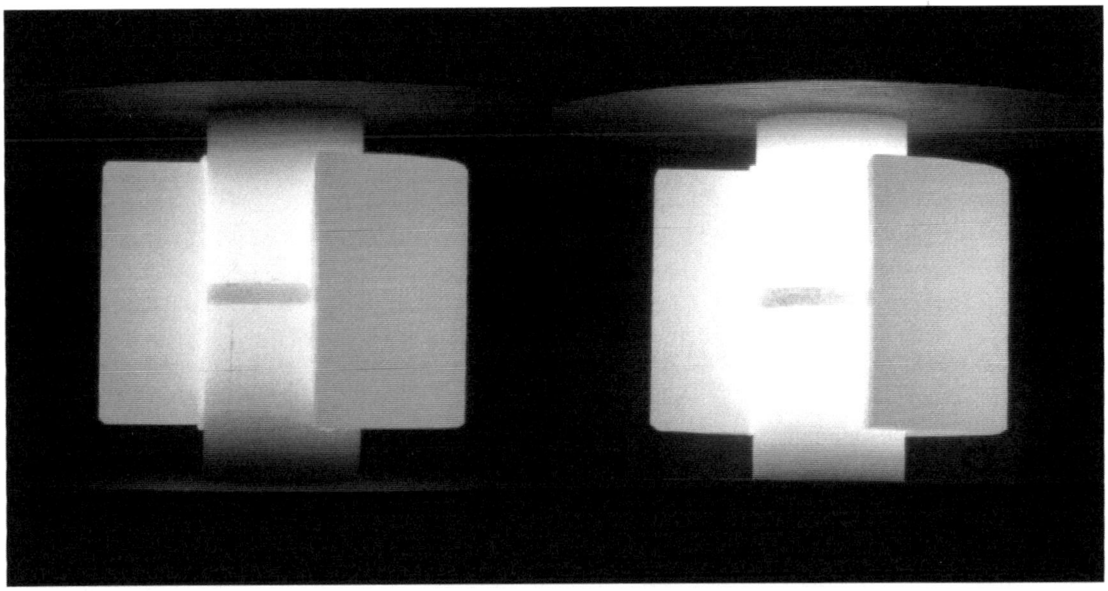

a *b*

Figure 3. Photographs of the press mold at high and low temperature: (a) surface temperature amounts 1100 °C, (b) surface temperature amounts to 1500 °C.

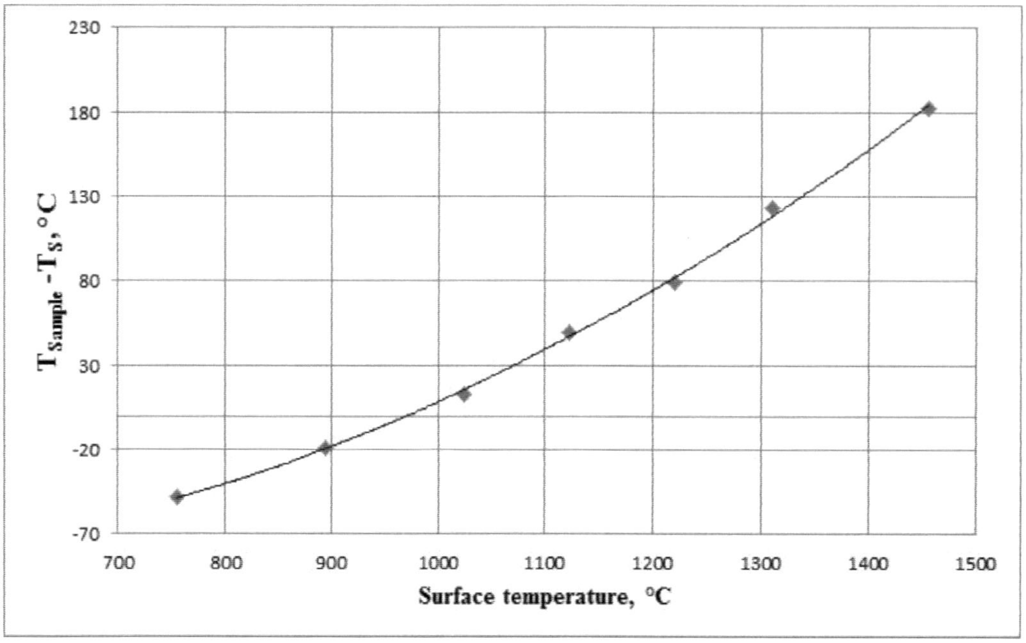

Figure 4. Graph of the dependence of the difference between the sample temperature and the surface temperature on the surface temperature

It was found that at the temperatures less than about 1000 °C the sample is colder than the surface In addition, there is a specific feature: the primary source of heat emission at temperatures about 1300°C - 1400 °C is the contact resistance at contact resistance at the punch-matrix borders. Then heat emission zone moves towards the punches. This can be explained by the fact that during the heating process the thermal expansion of punches occurs, and as a result the value of the contact resistance reduces.

4. Modeling

In addition to the parameters of materials used in the research, the modeling required determining the parameters of the thermal and electrical contact resistance at the punch-matrix border, as showed in Figure 5.

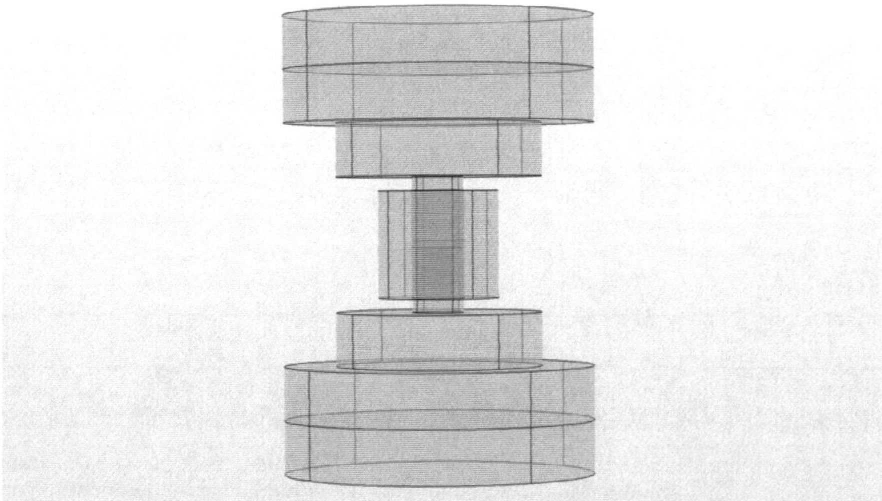

Figure 5. Contact resistance area.

IOP Conf. Series: Materials Science and Engineering **218** (2017) 012016 doi:10.1088/1757-899X/218/1/012016

The parameters of graphite, which the punches, press mold and spacers are composed of, as well as the parameters of aluminum oxide are taken from the research [1].

The reference experiments allowed choosing the temperature correspondence between the thermal and electrical resistance of the contact area of punch-matrix. The results of the simulation are presented in Figure 6.

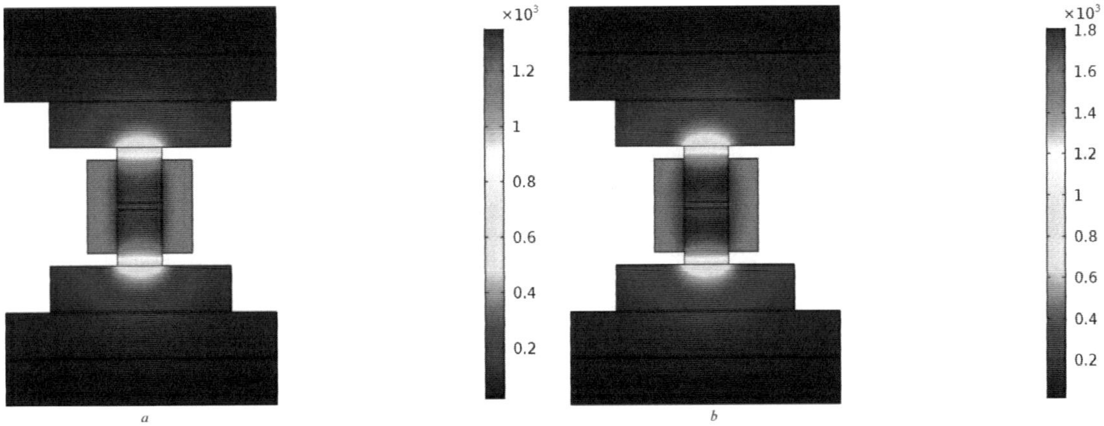

Figure 6. Results of modeling: (a) 1100 °C, (b) 1500 °C.

It is evident that the calculated data provide a good agreement with the experiment and reflect the changing of heat zone mechanism.

Since the actual mold is surrounded by a thermal insulation felt, it was necessary to adapt the results of modeling to this case, as shown in Figure 7.

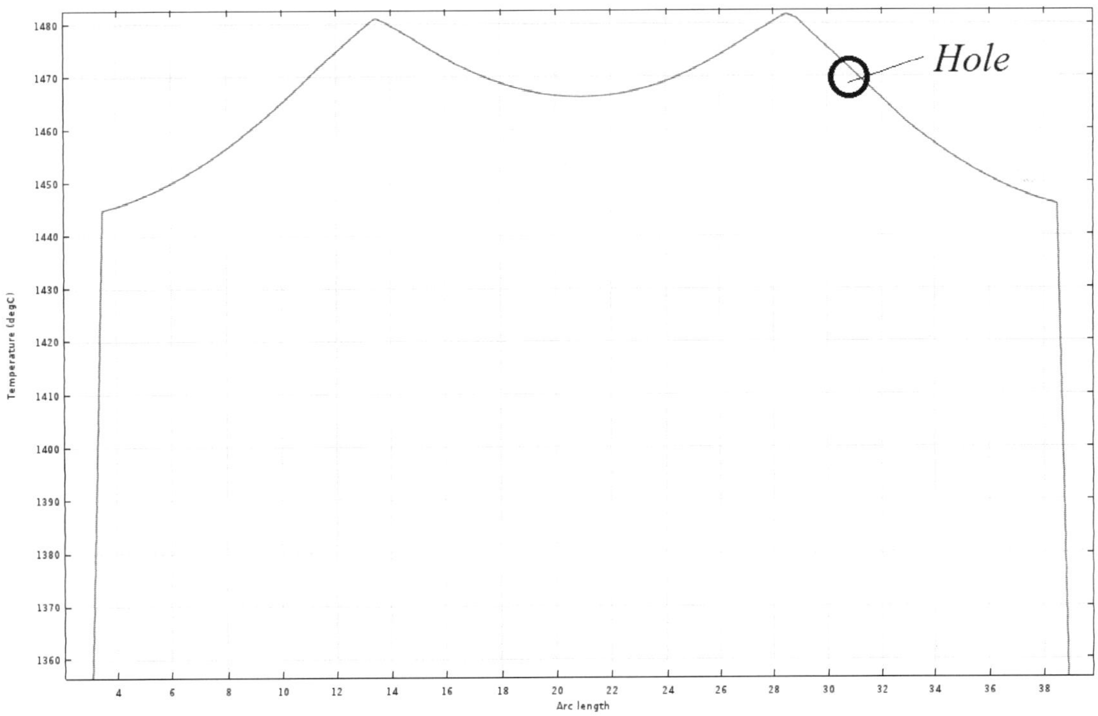

Figure 7. Temperature distribution in the diameter of the mold (the line passes through the sample) with thermal insulation.

The results show that with a good thermal insulation and with temperature measurement in an opening of sufficient depth (8 mm) the sample temperature can be controlled with a high degree of accuracy.

The model has also been generalized to the case of obtaining glazing beads, i.e. compacts with the height of 10 mm to 15 mm, the results shown in Figure 8.

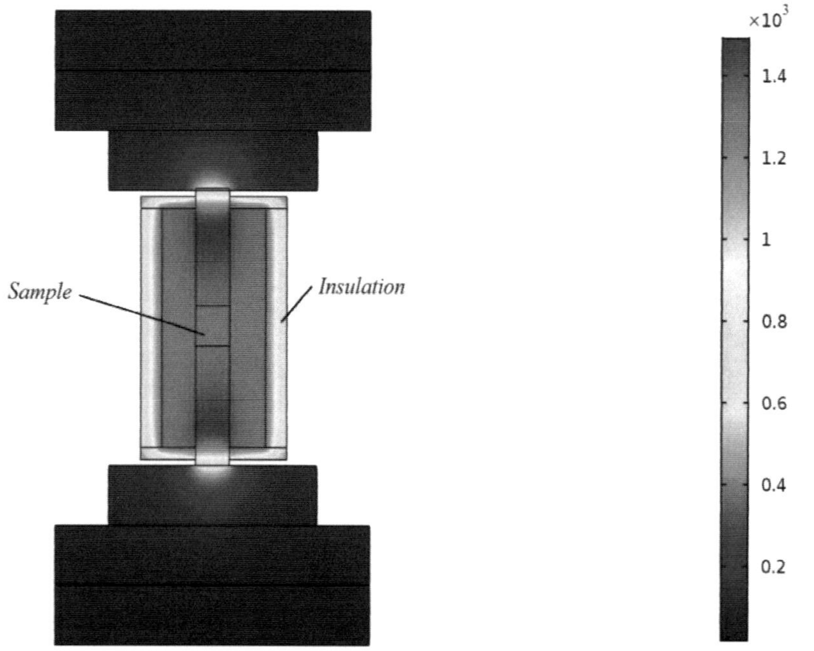

Figure 8. The temperature distribution in the diameter of the press mold for the case of producing glazing beads.

It appears that with a good thermal insulation it is possible to achieve a nearly uniform temperature distribution in the height of the sample. The difference of temperatures does not exceed 50°C to 70 °C. In this case, the Japanese machine model has an advantage over the German machine model, where the temperature is measured through an opening in the punch. Such measurement of temperature cause changes in the temperature field at the edge of the sample, which that does not allow achieving a uniform temperature distribution and therefore the uniformity of properties of the compact after sintering.

5. Discussion of results

The main feature of the experimental part of the research was a discovery that the main source of heat emission at temperatures less than 1300°C to 1400 °C is contact resistance at the punch-matrix borders. Then heat emission zone moves towards the punches. This can be explained by the fact that thermal expansion of the punches occurs during the heating process, and as a result, the value of the contact resistance reduces. Moreover, the resistance of the contact itself reduces with decrease of temperature.

Up to a certain point (1000 °C) the temperature of the sample is lower than the surface temperature. It is possible to assume that this relates to more intensive heat removal due to the emission and thermal conductivity (towards cooled electrodes) compared to its absorption in the sample. The calculation model showed a good agreement with the experiment, and its generalization to the real systems with thermal insulation will allow controlling the temperature of the sample by controlling the temperature of the surface or of the opening.

6. Conclusions

IOP Conf. Series: Materials Science and Engineering **218** (2017) 012016 doi:10.1088/1757-899X/218/1/012016

The experimental results allowed establishing the mechanism of change of the heat emission source, which moves from the contacts to the punches. Literature sources lack this information.

The temperature dependencies of electric and contact resistance on the punch-matrix border were selected. The calculation model based on these dependencies shows a good match to the experiment model.

The model was extended to the systems with thermal insulation and demonstrated the possibility of the sample temperature control by the means of controlling the temperature of the surface or the opening. In addition, the research allowed substantiating the possibility of producing glazing beads with a uniform temperature distribution in the sample height, and consequently the uniform distribution of properties. The results may be generalized to any ceramic materials and geometric features of the assembly structure.

The laboratory is financed by the Russian Government grant № 11.G34.31.0051 «Perspective technologies for producing new materials

References

[1] Manière C, Pavia A, Durand L, Chevallier G, Afanga K, Estournès C 2016 Finite-element modeling of the electro-thermal contacts in the spark plasma sintering process *J. Eur. Ceram. Soc.,* **36** 741–748

[2] Pavia A, Durand L, Ajustron F, Bley V, Chevallier G, Peigney A, Estournès C 2016 Electro-thermal measurements and finite element method simulations of a spark plasma sintering device *J. Mater. Process. Technol.* **213** 1327–1336

[3] Olevsky E A, Garcia-Cardona C, Bradbury W L, Haines C D, Martin D G, Kapoor D 2012 Fundamental aspects of spark plasma sintering II: finite element analysis of scalability *J. Am. Ceram. Soc.* **95** 2414–2422

[4] Maniere C, Pavia A, Durand L, Chevallier G, Bley V, Afanga K, Peigney A, Estournes C 2015 Pulse analysis and electric contact measurements in spark plasma sintering *Electr. Power Syst. Res.* **127** 307–313

[5] Munir Z A, Anselmi-Tamburini U, Ohyanagi M 2006 The effect of electric field and pressure on the synthesis and consolidation of materials: a review of the spark plasma sintering method *J. Mater. Sci.* **41** 763–777

[6] Giuntini D, Raethel J, Herrmann M, Michaelis A, Haines C D and Olevsky E A 2016 Spark plasma sintering novel tooling design: temperature uniformization during consolidation of silicon nitride powder *J. Ceram. Soc. Japan* **123** 403-414

Structure and properties of Al_2O_3 / Graphene nanocomposite processed by spark plasma sintering

V V Stolyarov [1, 2], A A Misochenko[1, 2], A G Zholnin[2], E G Grigiriev[2], and E A Klyatskina[3]

[1]Mechanical Engineering Research Institute of Russian Academy of Sciences 4, Maly Kharitonievsky line, 101990, Moscow, Russia
[2] National Research Nuclear University MEPhI (Moscow Engineering Physics Institute), Kashirskoye sh., 31, 115409, Moscow, Russia
[3]Instituto de Tecnología de Materiales, Universitat Politècnica de València, Camino de Vera s/n, 46022, Valencia, Spain

E-mail: vlstol@mail.ru

Abstract.The ultrasonic dispersion and spark plasma sintering technique have been used to process Al_2O_3 / Graphene nanocomposite (Al_2O_3 /G) with 0-2wt.% graphene. The topography of fracture, microstructure, density, microhardness, Young modulus, electrical and thermal conductivity of Al_2O_3 / G nanocomposite are investigated.

1. Introduction

It is known that the graphene, two-dimensional allotropic modification of carbon, possesses a number of unique physical and mechanical properties and can be used as the reinforcing additive for composite materials [1, 2]. It appeared that properties of such composites depend on material and matrix dispersion, and also an introduction method of a graphene. The composite on the basis of Al_2O_3 nanopowder and small adds of a multilayered graphene is of interest [3, 4]. At present the most successful for processing such composites is the spark plasma sintering (SPS). It is supposed that graphene flakes as the second phase will cause considerable wear reduction because of friction forces reduction on contact of nanometric and submicrometric structure elements [5]. However, data on tribological properties, and also heat conductivity and electrical conductivity are practically absent.

The purpose of work is demonstration of successful processing nanocomposite of Al_2O_3 /G by SPS technique and results of research of its physical and mechanical properties.

2. Experimental procedure

2.1. Materials preparation

To process a ceramic nanocomposite the mixture consisting of the nanopowder δ - Al_2O_3 (45 nm, IMET RAS, Moscow, Russia) and 0.5; 1 and 2 wt. % of graphene nanoflakes (δ = 3 nm, the lateral size of 10x10 μm) prepared by ultrasonic exfoliation method (3 x Graphene-tech, Zaragoza, Spain) has been prepared. After ultrasonic dispersion in solvent and the subsequent drying the samples with size of Ø15x2 mm have been compacted at 1550 °C within 10 min. under pressure of 50 MPa in vacuum SPS set (LABOX-625, SinterLand, Nagaoka city, Japan) as is described in [6].

2.2. Characterization

The bulk density of samples was measured by Archimedes method. Microhardness (HV) was measured on the FM-800 device along diameter at the load 3H during 10 s. Elastic modulus (E) was measured on the nanotester G200, MTS by a method [7]. Fractography researches and structure certification of graphene distribution are executed, respectively, on the electronic microscopes FESEM, ZEISS ULITRA 55, Oxford Instrument and Hitachi SU8000. Optimization of analytical measurements is performed within the approach described earlier [8, 9]. To evaluate the structural stability analysis of a graphene Raman spectroscopy was used at the room temperature by Renishaw via spectrometer with laser wave length λ=532 nm.

Temperature dependence of specific electroconductivity $\chi(T)$ has been determined by resistivity measurement on hot pressing set in Ar atmosphere (6.6 Pa) in the disc thickness direction under heating in the range of 20-1500°C with the speed of 20 °/min. Temperature dependence of heat conductivity of λ (T) was defined on the computer heat conductivity tester KIT-600 along a disk surface in the range of 20-195°C to within ±0.8 W/м*К. Tribological researches are executed on the Tribometer, CSM Instruments at indenter rotation according to the scheme «ball on disk», at load of 20 N and the speed of 10 cm /s during 15 hours. As a counter material the ruby ball of Ø 6 mm and HV 2400 was used. Sample surface was polished down to 1 μm and cleaned before and after the wear test. Wear track surfaces were examined by scanning electron microscopy. Wear rate (V_w) was calculated using the volume of a friction track (mm^3) which sizes were measured on the profilometer.

3. Experimental results

Density measurement results, microhardness and the elastic modulus of Al_2O_3/0.5wt.%G nanocomposite are given in the table 1. Density of a composite with graphene is slightly lower than the theoretical density of aluminum oxide. Microhardness and the elastic modulus of the composite with 0.5 wt. % graphene are 20-25% higher ones for the sample without graphene. Note that level of mechanical properties is higher, than for similar researches of micrometer composites [10, 11].

Table 1. Density and mechanical properties of Al_2O_3 sample with and without graphene

Graphene content (wt.%)	ρ (g/sm^3)	HV (GPa)	E (GPa)
0	3.99	20-24	380
0.5	3.95	27.4	456

The SEM photographs of nanocomposite fracture surface (fig. 1) show that graphene introduction promotes disappearance of microvoids (fig. 1a, b), at the same time slightly increases the size of matrix grains and leads to heterogeneity in graphene distribution.

Figure 1. SEM photographs of fracture surface for Al_2O_3/G composite: (a) 0%, (b) 0.5%; (c) 2%.

Increase in graphene contents up to 2% leads to formation of the submicron graphene agglomerates (fig. 1c) that was confirmed with the EDS analysis. It is important to emphasize that the initial

IOP Conf. Series: Materials Science and Engineering **218** (2017) 012017 doi:10.1088/1757-899X/218/1/012017

graphene flakes during of high-temperature SPS processing didn't degrade and remained stable that has been confirmed with Raman spectroscopy

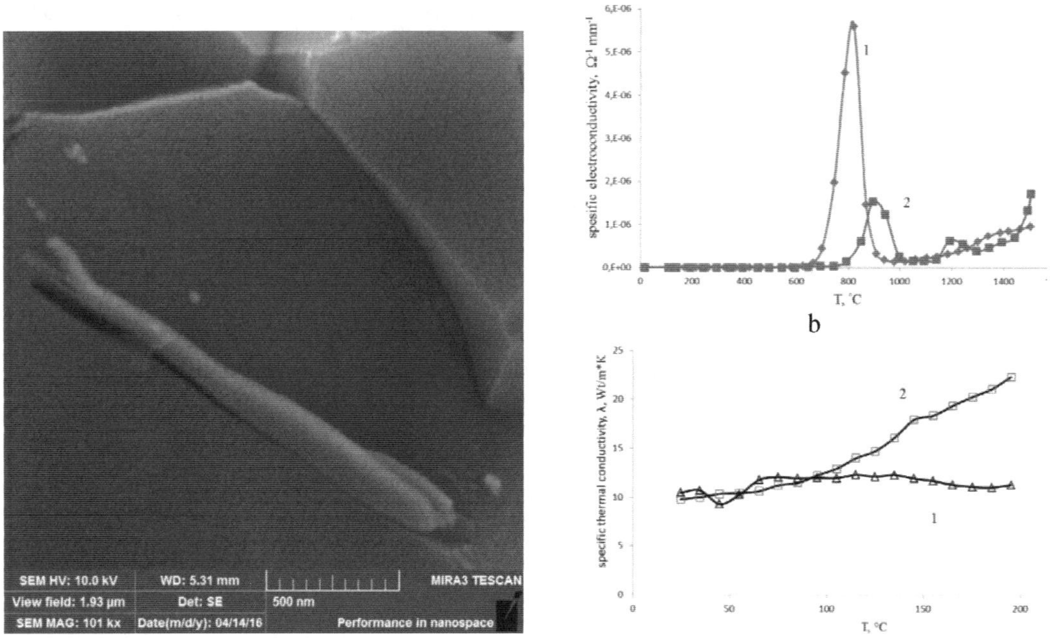

a c

Figure. 2 SEM photo of graphene flake (a) and temperature dependences of specific electrical conductivity (b) and specific thermal conductivity (c): 1 - Al_2O_3 / 2wt.%G; 2 - Al_2O_3.

It is rather difficult to find and investigate the graphene flakes morphology because of their small sizes and low contents. However it is possible to see that some graphene flakes can be placed inside Al_2O_3 grains in (fig. 2a). Nevertheless, some TEM data demonstrate that graphene particles can be located also on the grains boundaries [12].

The research of physical properties, for instance, electro-and heat conductivity of composites is of special interest (fig. 2b, c). First of all, the behavior of specific electroconductivity with temperature variations is in opposition relation with the one of metals. The same tendency is observed for the Al_2O_3 / 2wt.%G. Another feature for both materials is the manifested peaks of electrical conductivity in the area of 600 – 1000 °C. However samples with a graphene have shown stronger peak than a sample without graphene (fig. 2b). The peak of specific conductivity for a sample without graphene is weaker and is shifted towards high temperatures. The reason of peaks is perhaps connected with phase transformation of Al_2O_3 that demands an additional research. As for specific heat conductivity its value grows with a temperature for a sample without graphene and practically doesn't change for a sample with a graphene (fig. 2c). The last indicates a possibility of increase in heat-insulating properties at graphene introduction to composite material.

It is possible to suppose that tribological properties of Al_2O_3 composite will be depending on the graphene content. In fact, data in the table 2 demonstrate that wear rate V_w for samples with a graphene decreases by two orders in comparison with samples without graphene at insignificant reduction of friction coefficient

Table 2. Friction coefficients (K_f) and wear rate (V_w) of nanocomposite.

Graphene content (wt. %)	V_w (mm^3/h)	K_f
0	$5.4 \cdot 10^{-4}$	0.77
2	$< 2 \cdot 10^{-6}$	0.63

4. Conclusions

1. The method of graphene introduction to ceramics based on ultrasonic dispersion and spark plasma sintering can be successfully used for creation of high density and high-strength on Al_2O_3 composites.
2. Some surface properties as friction coefficient, microhardness, wear resistance are the most sensitive to the graphene content (to 2%), last of which raise to 30% and for 2 orders, respectively.

Acknowledgements

This work has been supported by the Russian Science Foundation N°16-19-10213. The authors acknowledge the Structure Investigation Department of Institute of Organic Chemistry, RAS for help with electron microscopy research

References

[1] Novoselov K. S. et al. 2005 *Nature* **438** DOI:10.1038/nature04233
[2] Stankovich S. et al. 2006 *Nature* **442** 282
[3] Kim H J et al. 2014 *Sci Rep* **4** 5176.
[4] Centeno A, Rocha V G, Alonso B, Fernández A et al 2013 *J Eur Ceram Soc* **33** 3201–3210
[5] Borrell A, Torrecillas R, Rocha V G, Fernández A et al 2012 *Wear* **274** 94-99
[6] Zholnin A G, Kovaleva I V, Yurlova M C, Ilina A M et al. 2015 *Phys Chem Mat Procs* **2** 73-79
[7] Oliver W C, Pharr G M 1992 *J Mater Res* **7** 1564-1583
[8] Kachala V. V., Khemchyan L. L., Kashin A. S., Orlov N. V., Grachev A. A, Zalesskiy S. S., Ananikov V. P. 2013 *Russ. Chem. Rev.* **82** 648 – 685
[9] Kashin A.S., Ananikov V.P. 2011 *Russ. Chem. Bull. Int. Ed.* **60** 2602 – 2607
[10] Benavente R, Pruna A, Borrell A, Salvador M D et al 2015 *Mater Res Bull* **64** 245–251
[11] Zholnin A G, Kovaleva I V, Rytenko V Yu, Pahilo-Daryal I O et al 2016 *Phys Chem Mat Procs* **1** 53-63
[12] Klyatskina E A, Grigoriev E G, Zholnin A G, Salvador M D et al. 2016 Structure and properties of alumina reinforcement with graphene nanoplatelets obtained by spark plasma sintering (unpublished)

Consolidation of materials by pulse-discharge processes

E L Strizhakov and S V Nescoromniy

Don State Technical University, Gagarin square 1, Rostov-on-Don, 344000, Russia

E-mail: nescoromniy@mail.ru

Abstract. The article presents the research and the analysis of the pulse-discharge processes of capacitor discharge sintering: CD Stud Welding, capacitor discharge percussion welding (CDPW), high-voltage capacitor welding with an inductive-dynamic drive (HVCW with IDD), pulse electric current sintering (PECS) of powders. The comparative analysis of the impact parameter is presented.

1. Introduction

Pulse discharge processes are widely used to produce permanent joints of heterogeneous materials. Common for them is the technological use of the pulse-discharge equipment allowing implementing the thermal and physical impact on the processed unit [1].

The block diagram of the pulse-discharge equipment includes a charging device, capacitive energy storage, a switching device and a technological block. Common for all the types of processing is a pulse current generator (PCG), and technological block includes the instrument and the equipment typical for ta particular type of processing (Fig.1).

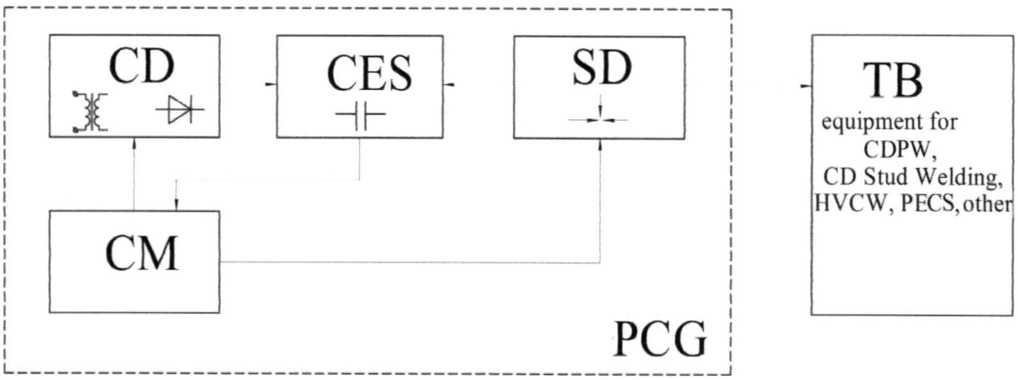

Figure 1. A block diagram of the pulse-discharge equipment: CD - charging device; CES - capacitive energy storage; SD - switching device; CM - control module; PCG - pulse current generator; TB - technological block; CDPW – capacitor-discharge percussion welding; CD Stud Welding - welding pulsed arc; HVCW – high-voltage capacitor welding; PECS - pulse electric current sintering of powders.

To improve the reliability of welded joints of rod elements made of non-ferrous metals with the main part of the metal structure, it is necessary to reduce the width of the fragile part by generating local heat and mechanically impacting the contact area. These requirements are met by the sintering methods with storing the accumulated energy in the capacitor banks: the CD stud welding technology, capasitor discharge percussion welding (CDPW) with a spring drive in air or in vacuum, high-voltage capacitor welding with an inductive-dynamic drive (HVCW with IDD).

Comparative analysis of rod elements welding methods allowed to summarize the area of their use [2-4].

2. Experimental results and disscution

The stud welding technology involves the starting of the impulce arc by a capacitor banks discharge due to the tearing off of the metal component from the flat part and further convergence of the surfaces caused by a press mechanism.The diameter of the welding rod elements does not exceed 12mm.

The following technological methods of the CD Stud Welding has been developed: stud welding by a discharge of the capacitor banks (CD), by a short electric arc with a short production cycle (SC), by an electric arc (ARC) [2]. The methods vary by the duration of the arc burning, $(1\div5)\cdot10^3$ ms, $(3\div500)\cdot10^3$ ms and $(100\div2000)\cdot10^3$ ms respectively.

In the SC and ARC methods, the end of the metal component has a tapered shape, and protective gases or flux fings are used for protection of the weld pool from the environment.

In the CD method the end of the rod element has a cylindrical protrusion.

Capasitor discharge percussion welding (CDPW) is used for the welding of the rod elements less than 6 mm in diameter. The process includes a capacitor banks discharge into the welding components and their convergence caused by a spring mechanism. During the discharge, an arc is forming between the surfaces and burning in the metal fumes. The duration of the burning depends on difference between pressures of metal fumes and of spring mechanism. When the sping drive overcomes the metal fumes pressure, the rod element sinks into the molted weld pool, simultaneous crystallization occurs and the weld joint is formed in the liquid phase.

Experimental studies of CDPW in vacuum, with the pressue of aluminium components to steel sheets of 0.133 Pa to 0.00133 Pa showed that duration of arc burning increased at constant current and equals 3 ms to 20 ms [2,3]. The increase in duration leads to intensification and decrease of matal fumes pressure comparing to air enviroment, reduction in rod element damping against the metal plate and larger displacement of metal from the welding zone.

The process of high-voltage capacitor discharge percussion welding with an inductive-dynamic drive (HVCW with IDD) has been developed for welding of the rod elements to the main parts in homogeneous and heterogeneous combinations with diameter of 6 mm to 25 mm. [4] The scheme of the process is presented in Fig. 2.

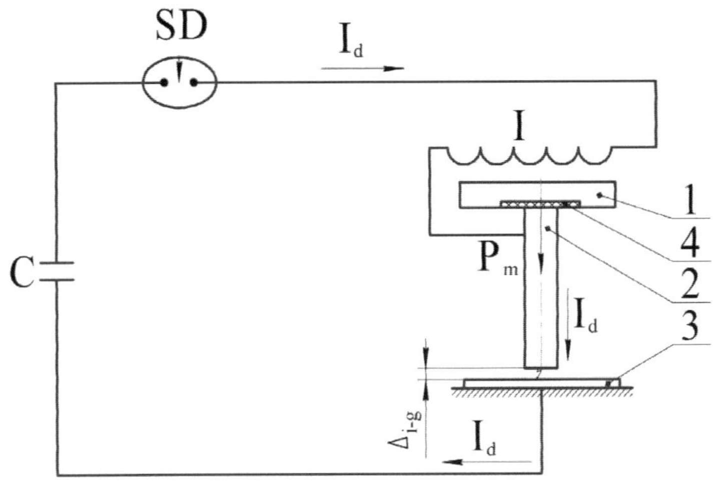

Figure 2. The discharge circuit of the HVCW device with IDD
1 – pusher; 2 – rod element;
3 – plate (main part);
4 – dielectric insertion;
I_d – a discharge current;
P_M – magnetic pressure;
SD - switching device;
C - capacitive energy storage,
$\Delta_{i\text{-}g}$ – initial gap.

IOP Conf. Series: Materials Science and Engineering **218** (2017) 012018 doi:10.1088/1757-899X/218/1/012018

During the discharge of the capacitor banks the current Ip flowing down the coils of the inductor I generates an alternating magnetic field which induces eddy currents in the pusher 1. Interaction of the eddy currents in the pusher 1 with the magnetic field of the inductor leads to generation of the magnetic pressure Pm, and so the mechanical impact on the rod element 2 occurs.

An inductive-dynamic drive (IDD) [1,4] is connected in series with the welded parts 2 and 3 into the generator circuit in order to synchronize the mechanical and thermal pulse loads.

IDD consists of a flat inductor with a pusher in the working area of it. Duration of the current discharge depends on the number of turns of the inductor and amounts to 100 to 500 ms. Comparative analysis of dependences of the welding current on the time of its occurrence in different capasitor discharge sintering methods is presented in Figure 3.

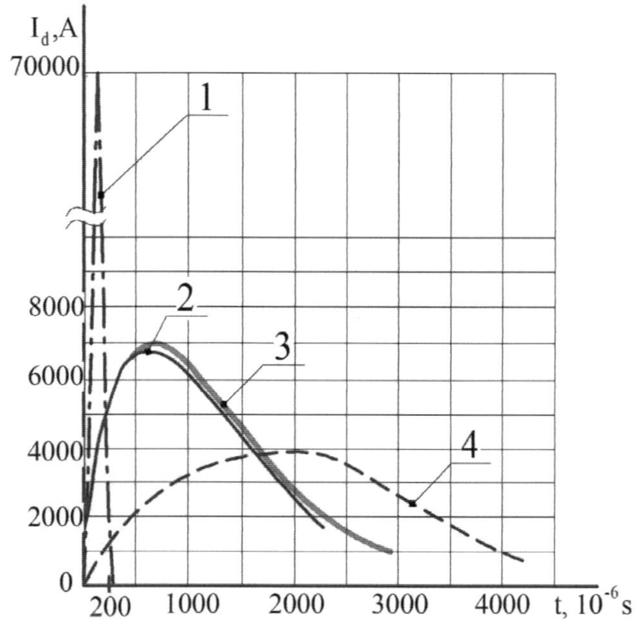

Figure 3. Graph of I=f(t) stud d=6mm alloy AMg3 with steel sheets:
1 - HVCW device with IDD;
2 – CDPW in the atmosphere;
3 – CDPW in vacuum;
4 – CD Stud Welding.

The dependences I=f(t) 2, 3, 4 relate to the strict parameters of the modes of resistance welding exposure.

Using for the HVCW with IDD process the high-frequency pulse current generators with natural frequency (f) of 100 kHz to 200 kHz and forging force of the IDD (Pm) about 105 N/mm^2 to 106 N/mm^2 allows super-rigid processing modes, position 1.

The macrostructure of welded joints of aluminium rods with steel plates is shown in Fig. 4.

The thickness of the crystallized metal in the joint area is less in result of CDPW in vacuum than in result of CDPW in the atmosphere (Fig. 4 a,b), as evidenced by studies [3].

The width of the crystallized metal area formed with the use of CD Stud Welding technology occupies an intermediate position comparing to that of CDPW in the atmosphere and CDPW in vacuum (Fig. 4, c).

In the HVCW with IDD process there is no crystallized metal in the joint area due to a small amount of molten metal and its total displacement out of the welding zone due to the super-rigid welding mode.

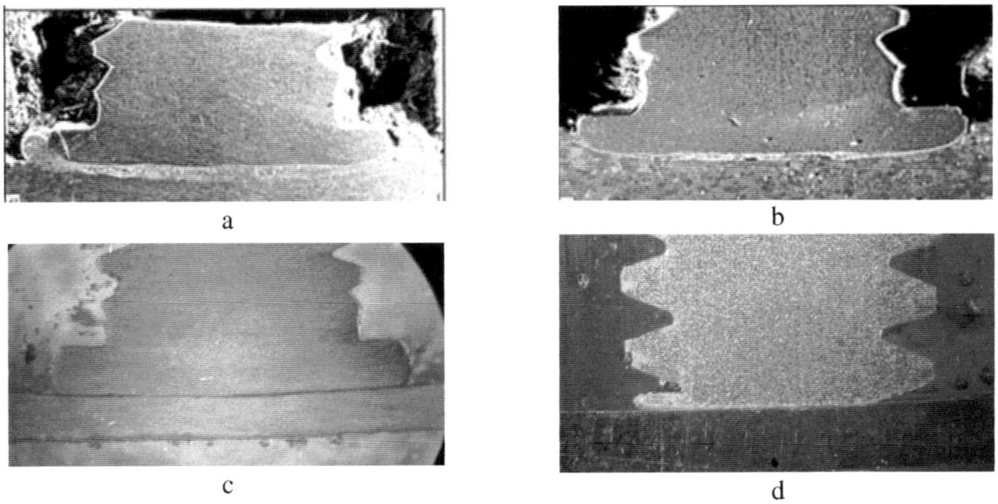

Figure 4. Macrostructure of welded joints of aluminium rods with steel substrate (x16):
(a) CDPW in the atmosphere; (b) CDPW in vacuum (0,00133 Pa); (c) CD Stud Welding;
(d) HVCW device with IDD.

Figure 5 presents the microstructure of welding joint of M1 copper alloy C01 with B63 alloy. It has been established that the border between the joining areas of dissimilar alloys is typical for high-speed welding of metals in the solid phase; there is a distinct line of border – the area of "setting". Microhardness in the welding zone is elevated, which is typical for pulse welding methods such as magnetic pulse welding and explosion welding.

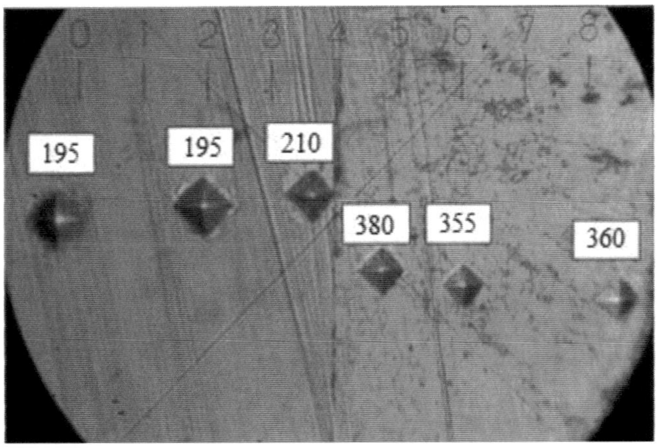

Figure 5. Microstructure of a zone of welding joint of M1 copper alloy C01 with an alloy B63, x485.

By changing the value of initial gap Δi-g, the inductance of a plane inductor and the mass of the IDD pusher, it is possible to adjust the duration of the heat W_h and force W_m impact. Analysis of the experimental data showed that the maximum strength of welded joints is at the ratio of energy for heating and mechanical impact W_h/W_m equal 2/3. When the W_h/W_m ratio is less than 2/3 there are pollution and lack of fusion observed in the welding zone. When W_h/W_m ratio is higher than 2/3, the joint area is formed by overheated and crystallized metal.

HVCW devices with IDD are successfully used for pulse electric current sintering (PECS) of powders, particularly for the welding of the output circuit of electrochemical protection, Figure 6 [5-7].

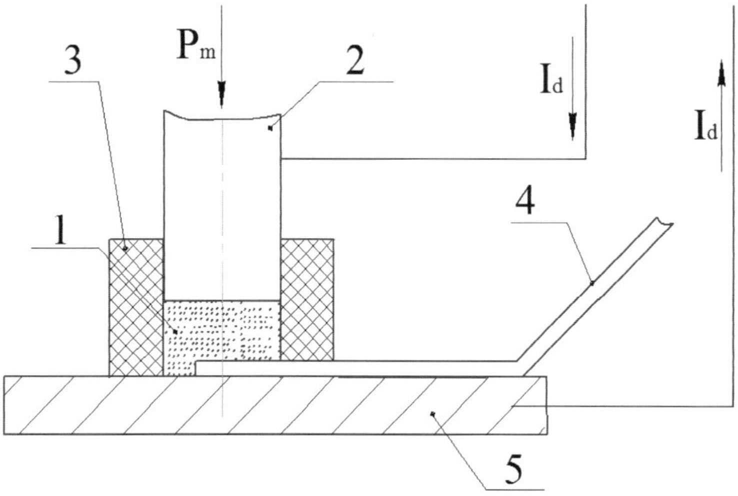

Figure 6. The welding of the output circuit of electrochemical protection 1 – powder; 2 – electrode punch; 3 – dielectric matrix; 4 - wire of electrochemical protection; 5 – the body of the pipe; I_d – discharge current; P_m – magnetic pressure

In order to create powerful current pulses the battery of high-voltage pulse condensers with capacity of 150 μF to 600 μF is used, which provides intensive energy discharge through the electrode punch 2 in the powder composition 1. Connection of an IDD in series to the electrode-powder-pipe system provides the synchronization of dynamic pressure Pm with the thermal effect I_p, which allows increasing the density of the powder composition [6, 7].

The input energy spent on the thermal and physical impact amounts to 3000 to 7000 J. The duration of the current I_p flowing through the electrode punches and through the powder does not exceed 300ms.

References

[1] Nescoromny S V and Strizhakov E L 2012 Special methods of magnetic pulse welding *LAP LAMBERT Academic Publisching* 152
[2] Kaleko D M 2013 The welding of steel studs to aluminum *Automatic Welding* **8** 38-43.
[3] Paton B E 2012 Condenser stud welding in vacuum *Automatic Welding* **4** 42-47.
[4] Strizhakov E L 2015 Classification of methods and examination of the process of high-voltage pulse electrostatic welding *Welding production* **3** 42-46
[5] Grigoriev E 2007 Electro discpaction of WC – Co composite material containing particles of diamond *Materials Science Forum* **534–536** 1181–1184
[6] Lemeshev S V 2015 Technology of joining dissimilar materials electric pulse welding-pressing *IV Conference on Science and Technology* 442-444
[7] Strizhakov E L Nescoromny S V and Rachkov M A 2014 Device for electrostatic percussion welding with magnetic pulse drive Patent Russian Federation N 96515, B22F 3/087

Combined effect of the electric current magnetic field and microwave radiation during the deformation of the stainless steel

O A Troitskiy and V I Stashenko

Mechanical Engineering Research Institute named after A.A.Blagonravov of the Russian Academy of Sciences, Maly Kharitonyevsky Pereulok, 4, Moscow, 101990, Russia

Abstract. The work determines the influence of microwave radiation on the processes of active deformation and relaxation of mechanical stresses in loaded samples of stainless steel under the effect of current pulses and longitudinal and transverse orientation of the electric-field vector E of the microwave radiation to the axis of the deformated sample. With the longitudinal orientation of the vector E of the microwave radiation and under the action of the current the effect of metal softening increased from 22% to 30%. The analysis of the microstructure of the samples showed a significant influence of external energy impacts on the deformation of steel grains.

1. Introduction

The passage of short current pulses of high density about 10^{3} A/mm^2 and length about ~ 0^{-4} sec cause a reduction of deformation stress in metals by tens of percent [1,2], and according to data [3] obtained by the double etching method, there is a consequent increase in mobility of dislocations caused by effect of electric current. The action of the electroplastic effect (EPE) still provides a discussion, however, the results of plastic deformation facilitation and improved mechanical properties of metals have been used successfully in the various technologies of metals and alloys chipless shaping. It was theoretically predicted by M. Molotskii [4] that spin softening of metal is possible during the electroplastic deformation. Molotskii suggested conducting a detailed research of metallic samples containing a variety of paramagnetic admixtures.

It should be noted that the experimental proof of the existence of the resonant alteration of plastic properties of crystals in mutually perpendicular constant and microwave magnetic fields under the electron paramagnetic resonance was first obtained by Golovin and Margunov in [5].

Implementation of the spin softening of metal due to the effect of the electron paramagnetic resonance on the mechanical properties requires the following conditions:

- magnetic fields (MF): intrinsic magnetic field of pulse current as it passes through the metal sample, and electromagnetic microwave radiation covering the sample must be crossed;
- thermodynamically nonequilibrium processes must be present in the samples, particulary the generation of fresh dislocations which is provided by active deformation of samples during constant speed strain or the intermittent mode alternating strain with pauses of stress relaxation without removal of the load.

The purpose of the research was to study the effect of microwave radiation on electroplastic deformation of metal.

2. Experiment method and materials

Stainless steel 12X18H10T was chosen for the study of the effect of microwave radiation on the mechanical properties of metals as an adequate object for research of EPE. Steel is composed of ferromagnetic iron and nickel, as well as of paramagnetic impurities, which have their own magnetic moment and under the influence of external fields (under the effect of the electric current magnetic field) orient on the field, creating a resultant field greater than the external fields. The 12X18H10T steel contains titanium which is a paramagnetic metal, and its paramagnetic properties increase with increasing temperature, which is partially occurs during the electroplastic deformation of metal. These facts allow considering the 12X18H10T steel a satisfactory object.

The 2,45GGts frequency used in the microwave ovens, was chosen for the study of the effect of microwave radiation on the mechanical properties of metals in plastic deformation by strain with simultaneous impact of pulsed current [6]. Transmission of electromagnetic (EM) radiation from a microwave radiation source - a magnetron - to the deformed sample was carried out with a rectangular waveguide (see Fig.1).

Figure 1. Tested sample of stainless steel 12X18H10T with thickness equals 0,2 mm, length of the working part equals 28 mm in isolated grips with current leads inside the waveguide, the magneton in underneath. EM field *E* is directed across the sample axis. The tests are conducted in a horizontal rupture testing machine IR 5047-50-10

EM energy was emitted by the magnetron with an antenna in the form of electrical pin. The pin entering the waveguide was located at a distance of about a quarter of wavelength λ. One side of the rectangular waveguide was short-circuited by the metal wall. The pin of the magnetron emitted electromagnetic waves (EMW) in all directions - straight waves toward the load (of deformated sample) and return waves in the opposite direction toward the blind wall of the waveguide, which were folded after the reflection. The relation $b/a < 0,5$ must be satisfied for the standard waveguides. At $\lambda/a < 2$ the only type of waves propagating on the waveguide is H_{10}, which are recommended for use. The sides dimensions *a* and *b* can be determined according to wavelength. [6]. The distance from the pin to the axis of the deformated sample was a multiple of wavelength λ. In this case, the electric field strength on the sample is at maximum value. At the distance of $\lambda/4$ behind the sample the waveguide was confined by a blind wall. The length of the waveguide from one blind wall to the other equaled $\lambda + \lambda/2$. The waveguide box was made of stainless steel with slots in mutually perpendicular planes for placing the test sample. Remote control of the magneton was used in the experiments. The orientation of electric-field vector *E* of microwave radiation was varied by turning the waveguide box on 90^0 to the deformated sample. The temperature of the samples was measured with a thermocouple during the microwave radiation and and single current pulses. It didn't exceed 60^0C to 80^0C.

3. Results of the experiment

Effect of microwave radiation on the plastic deformation of metal was studied in two sets of the experiment: 1) with the active strain at different values of constant speed; 2) with the stress relaxations.

At the active deformation of the sample, single pulses were first submitted with the current density about 1000 A/mm^2 and duration of 250 μs. Then, after increasing the load by 50 N, the current pulse of the same value and microwave radiation were submitted. Figure 2 shows a fragment of machine diagram of sample deformation σ - *u*, where σ – conventional stresses, *u* – movement of the machine grip. The diagram shows the alteration between sharp decreases of deformation stress (stress surges caused by current pulses leading to the softening of the metal) and surges of current and microwave radiation. Field E orientation of microwave radiation was *transverse*. The electroplastic effect of pulse current on plastic deformation of the stainless steel increased in the presence of microwave radiation. Increased surges of stress reduction during the microwave radiation indicate the additional radiation exposure to the active metal deformation. A slight thermal effect of the current and the microwave did not cause power surges.

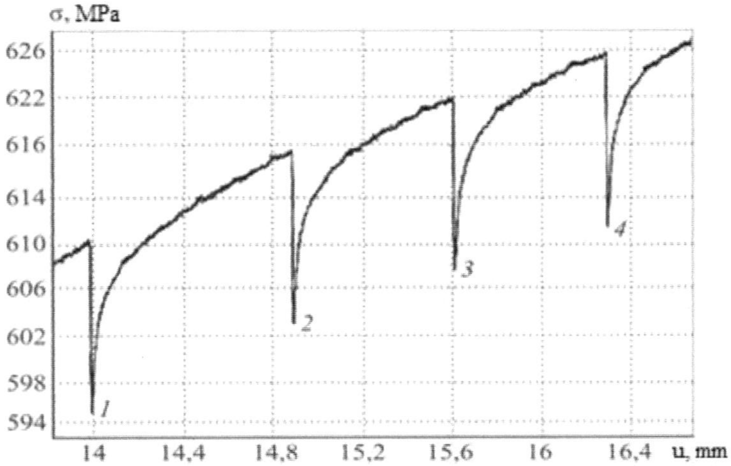

Figure 2. Diagram of sample deformation with surges of stress at the current pulses impact (1,3), current and microwave radiation during the active deformation (2,4). The value of surges: 1 - 14,0 MPa;
2 - 14,8 MPa; 3 - 14,3 MPa; 4 - 15,0 MPa.

Figure 3 shows the data measurement of softening spikes for five samples under the action of current pulses of high-density of about 1000 A/mm^2 (the lower area confined by the dashed lines) . Upper data (confined by the dashed lines) is for the same samples under the simultaneous current pulses and microwave radiation. The orientation of the electric field E is transverse.

The second set of experiment included relaxation of mechanical stresses during short stops in straining of samples with 3-minute pauses without unloading. The value of stress relaxation at different variants of internal and external energy influence was measured. Strains before each new series of measurements started with a higher load up to the sample failure. Each new series of measurements started with the load level 50 N to100 N higher than the former. Each series of measurements consisted of four pauses: a) without any influence; b) with microwave radiation at different orientations of vector E; c) with current pulses; г) with a combined action of microwave radiation and current pulses.

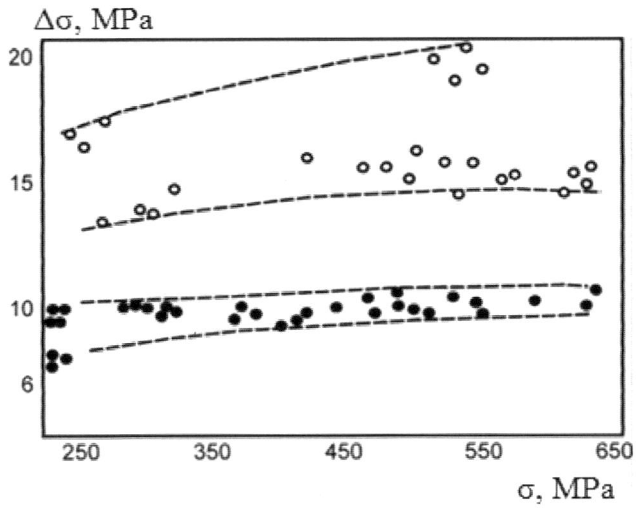

Figure 3. The dependence of the stress surges Δσ on the value of applied stress σ

It was possible to conduct 12 to 16 measurements on a sample. When the load amount reached 1400 - 1500 H a failure in the necking of samples usually occurred. During the stress drop at first 2-3 seconds after stop of active deformation without any action (a), relaxation amounted no more than 5.3%; with microwave radiation (b) and the pulse current (c) is amounted to 4.5% and 7.6% respectively. In the (d) setting – with a combined action – taking into account the substraction of the thermal action – the effect amounted to 12% to 13%. The field E orientation was *transverse*.

Stress drop during the 3-minute relaxation break amounted to 8%, with the action of the microwave radiation and current it amounted to 7.5-12.3% and 10-13% respectively. In all cases, the stress drop occurred with an increase in the level of applied stresses. Concurrently the combined effect of current and microwave radiation provides a relaxation amount about 14.6-22.5%. Electroplastic effect of pulsed current [1,2] on the plastic deformation of the stainless steel amplified in the presence of a microwave radiation. Stress reduction due to heating of the sample was subtracted from the effect of exposure. Dependence of the observed effect on speed has been studied. The effect of microwave radiation increases with increasing deformation rate that can be attributed to an increase in the number of unfinished displacements and moving dislocations under the influence of microwave irradiation, facilitating their discharge. The experiments allowed to establish previously observed phenomenon of the disappearance of the magnetic properties [2] of stainless steel at electroplastic deformation by suppressing the austenite-martensite transformation. As a result of usual transformation the necking of the samples becomes magnetic because of surges alpha-phase over-stress or martensite deformation, which has magnetic properties. This phenomenon was first registered during microwave radiation impact on deformated metal. Table 1 shows the results of stress drop for the sample, which in the process of stress relaxation was effected by the field vector E axially to the test sample. As seen from these data, the effect of the joint action of 30%. As seen from the data the combined effect amounts to 30%.

Preliminary analysis of the microstructure of the samples showed a significant dependence of external energy impacts on stainless steel.

Area and perimeter of grains decrease under the influence of electric current and microwave radiation, i.e. shattering of grain occurs and microstructure of the deformated part of the sample becomes fine-grained, the grains take a more rounded shape, with a predominant axial orientation of the structure, the length, width and elongation of the grains substantially reduce, as well as vertical and horizontal projections of grains. Equivalent diameter and the average grain size, as well as the Martin diameter, are reduced by effect of external energy impacts.

Electroplastic deformation is a complicated multistage process [2]. Data on the stainless steel microstructure obtained after the energy impacts and achieved significant deformations of the samples indicate the manifestation of the inverse Hall-Petch relation. The Hall-Petch law gives a quantitative description of the polycrystalline material yield strength increase with the decrease of grain size.

Table 1. The amount of load reduction during the stops the machine gear for a full pause of stress relaxation for 3 minutes for the different types of test

Action	Measuring ranges of applied force *F* and stress surge ΔF/F	
	F (kN)	ΔF/F (%)
No action	1-1.45	8.6-8.8
Current pulses	1-1.45	16.8-17.5
The E field is *transverse:*		
- microwave radiation	1-1.45	11.2-11.4
- current pulse and microwave radiation	1-1.45	21.9-22.5
The E field is *axial*		
- microwave radiation	1-1.45	11.9-12.3
- current pulse and microwave radiation	1-1.45	29.8-30.0

The basis of this dependence is the dislocation mechanism of plastic deformation: grain boundaries impede dislocation motion. Apparently, the predominant deformation mechanism changes with increasing deformation under the impact of pulsed current and microwave radiation. The mechanisms of reverse Hall-Petch effect is currently insufficiently studied [7].

4.Conclusion

The research experimentally determined for the first time the effect of microwave radiation on the process of plastic deformation of metal: with the active sample deformation by strain and relaxation of mechanical stresses, with the orientation of field vector E of microwave radiation axially or transversely. With the *axial* orientation of the microwave radiation vector E and action of current pulses the effect of softening of metal increased by 8% (from 22% to 30%). The dilatometric effect of the sample lengthening due to heated by the current and microwave radiation was taken into account by measuring the decrease in deforming force during the relaxation of mechanical stresses. The value of the pinch effect of current pulses in flat thin samples is low because it is proportional to the cross-sectional area of the sample [8]. The results of this study suggest the presence of additional deformation mechanism during the impact of pulse current and microwave radiation and indicate the possible effect of electron paramagnetic resonance in the structural defects of the deformed metal, which lead to a change in its plastic properties by the mechanism of the spin softening of metal. The results of the research can be used for any type of metal forming by pressure for intensification of the plastic deformation of the metal.

References

[1] Troitskiy O A 1969 Electromechanical effect in metals *Pisma v ZhETF [Letters to the Journal of Experimental and Theoretical Physics]* **10** 18-22
[2] Baranov Iu V, Troitskiy O A, Avramov Iu S, Shlyapin A D 2001 Physical basis of electropulse and electroplastic treatments and new materials (Moscow: MGIU) 843
[3] Zuev L B, Gromov V E, Kurilova I F 1978 Mobility of dislocations in zinc single crystals under the effect of current pulses *Academy of Sciences of the USSR* **239** 84-87
[4] Molotskii M 2003 *Philosophical Magazine* **83 (12)** 1421
[5] Golovin I I and Morgunov R B 1999 *The Journal of Experimental and Theoretical Physics* **115 (2)** 605–624
[6] Watson H 1972 Microwave semiconductor devices and their application (Moscow: Mir) 660
[7] Carlton C E and Ferreira P J 2007 What is behind the inverse Hall-Petch effect in nanocrystalline materials? *Acta Materialia* **55** 3749-3756
[8] Troitskiy O A 1977 Plastic deformation of metal, caused by the pinch-effect *Bulletin of the Academy of Sciences of the USSR. Physics Series* **6** 118-122

IOP Conf. Series: Materials Science and Engineering **218** (2017) 012020 doi:10.1088/1757-899X/218/1/012020

Generation of mechanical vibrations in metal samples by the use of the pinch effect

O A Troitskiy[1], O B Skvortsov[1,2] and V I Stashenko[1]

[1]Mechanical Engineering Research Institute named after A.A.Blagonravov of the Russian Academy of Sciences, Maly Kharitonyevsky Pereulok, 4, Moscow, 101990, Russia

[2] RDC Balancing Machine Plant LLC, Varshavskoe shosse, 46, Moscow, 115230, Russia

Abstract. The article presents the recent research in electrodynamic processes for metal samples exposed to current pulses. The pinch effect and the skin effect cause the vibration of the metal rods. The results of these studies show how current and magnetic field interact with material samples of gold, silver and copper. The analysis allowed establishing the dependences of peak acceleration on current density and conductor diameter. The dependencies can be used in metal workings and for nondestructive testing.

1. Introduction

The influence of large pulse currents on the mechanical condition of metal samples, leading to the disintegration of the conductor to the number of fragments has been described over fifty years ago [1]. To a large extent, such processes involve the interaction of the current-carrying conductor with its own magnetic field and with considerable mechanical stresses generated concurrently [2]. The static forces exertion on the sample material may be described as the influence of the pinch effect [3-5]. Change of mechanical properties during the metalwork caused by current became known as electro-elasticity phenomenon [3]. In recent years a number of research papers has been published studying the changes of the properties of metal conductors caused by transmitting pulse currents through them [6-10]. The authors offer different descriptions of the physical processes occurring in metal samples under the influence of pulsed currents.

2. Physical processes in the conductor carrying the pulse current

Transmission of the current pulse though conductor is accompanied by electromagnetic effect, as well as heating effect. In order to reduce the influence of temperature changes due to the metal heating during the experiments it is preferable to reduce the duration of the current pulse. However, with short pulses of current in the conductor the influence of the skin effect also becomes significant. In these conditions it is not the value of magnetic induction that is important, but the time derivative of it. The change in the magnetic field causes the current distribution in the surface layer of the conductor. This does not imply the absence of heat generation in the fields of electric current. The destruction of thin wires when exposed to short pulses was examined in [6]. The released heat may cause significant dynamic loads. The authors note that such current may cause an explosive destruction of the skin layer of the conductor.

The transmission of pulse current can also lead to a change in the phase composition of the conductor material [7], which is accompanied by its deformation. The combined effect of static load and pulse current can also be accompanied by structural changes in the material [8].

Since the deformation of the samples occurs at a relatively low heat generation and relatively small currents, mechanical deformations may be caused by forces of electromagnetic field. The research papers [9-11] examine the mechanical impact of the intrinsic magnetic field on the sample carrying the pulse current. The authors establish that mechanical vibrations of the sample are concurrently determined by magnetohydrodynamic processes. It is of interest to consider the features of the accompanying processes.

3. The method of experiments

Pulse current was passed through cylindrical samples of various metals (gold, silver, copper) with different diameters. The experimental setup is shown in Figure 1. In order to reduce the thermal effect the current pulses had a short duration - about 150 microseconds. Varying the amount of current was possible by changing the voltage. The transmission of current was controlled by a magnetic field sensor MLX90242ESE-GDA-000. In order to manage the sensitivity, the distance between the sensor and the sample was chosen to equal 4 mm. Before conducting the test the magnetic field detector was calibrated by transmitting standard current pulses. Characteristics of the magnetic field sensor provides linearity no worse than 1% in the range of magnetic induction of ±0,15 mT

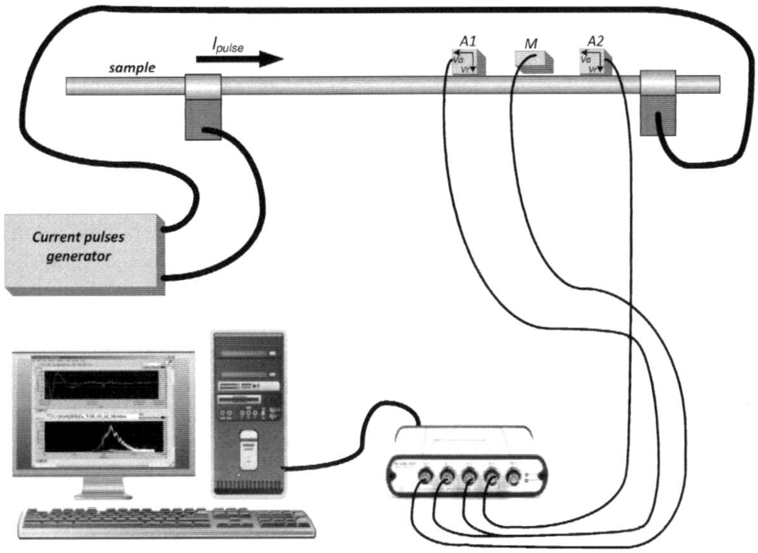

Figure 1. The structure of the measuring stand.

Control of vibration during the transmission of current pulse was carried by three-component piezoelectric accelerometers AP20 of small size and weight. Such accelerometers allow controlling vibration signals in the band up to 40 kHz. Collecting and processing of measurement data was conducted by analog signal acquisition unit 4431 NI USB and by software complex LabVIEW [12].

4. Results of the experiments

The silver samples had the lowest inherent resistance, which allowed setting the amplitude of passed current in the widest range. Assessment of the radical acceleration values characterizing the dynamic forces on the surface of the samples shows a relation close to a quadratic, a typical manifestation of the pinch effect in metals. The corresponding graph is shown in Figure 2.

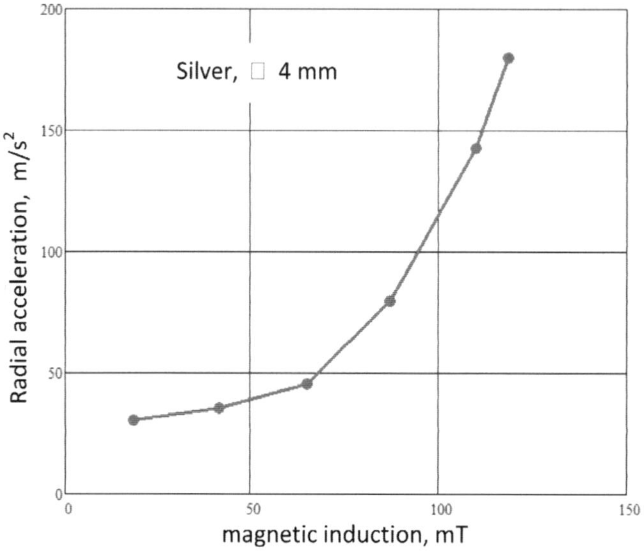

Figure 2. Correlation between radical acceleration and magnetic induction value. Silver sample, 4 mm in diameter.

The response of the sample to the transmission of the current pulse can be divided into two stages. In the first stage, during the current pulse transmission the excitation of deformations and redistribution of current occurs under the influence of change in the magnetic field generated by the current. During the second stage there are dampening intrinsic vibrations in the sample. Such vibrations may be used for providing nondestructive testing [13].

To analyze the processes occurring during pulse current transmission it is most interesting to measure extensively the signals in the first time stage, during the action of the magnetic field. Figure 3. shows the dependence of the magnetic field and the changes of vibration on time. The acceleration of the surface layers of the sample reaches the highest values in the beginning and in the end of the action of pulse current.

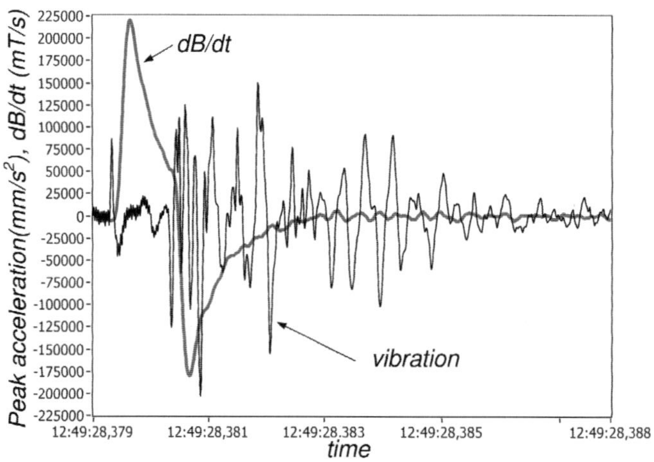

Figure 3. Simultaneous recording of the derivative of the magnetic induction and peak acceleration V2r.

The assessment of the displacement of surface layers is obtained by double integration of the acceleration signals in the axial and radial direction. It should be noted that different signals from different directions of various vibration sensors at the stage of active action of current are characterized by correlation (correlation coefficient more than 0.8). At the same time, the signals from the same direction and various sensors are poorly correlated (correlation coefficient of less than 0.3). This is also confirmed by the form of signals shown in Figures 4 and 5.

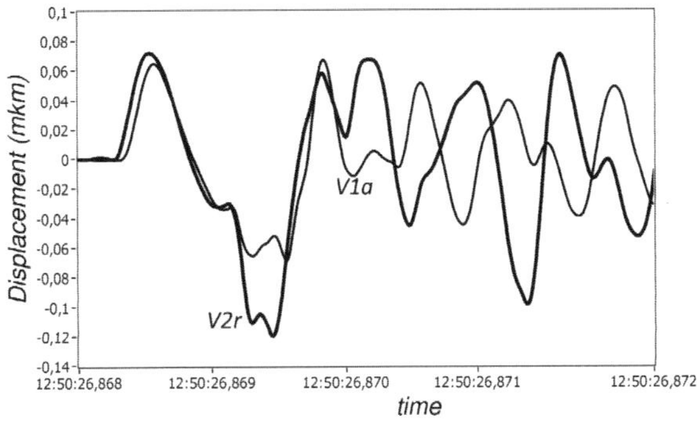

Figure 4. Simultaneous recording of axial vibration measured by a first *accelerometer* V1a and of radial vibration measured by a second accelerometer V2r.

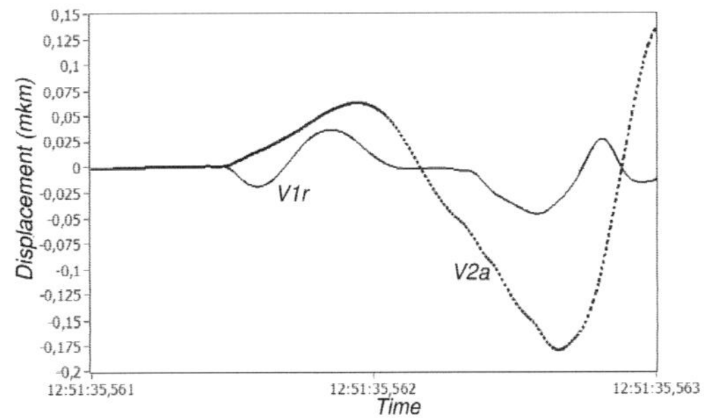

Figure 5. Simultaneous recording. Simultaneous recording of accelerations during the active action of the current pulse.

Such dependences may be regarded as a confirmation of the turbulent nature of MHD processes which cause mechanical vibrations during transmission of pulse current.

Figures 6 and 7 show the dependence of the peak value of the acceleration on the peak current value for silver and gold samples of different diameters. With an equal current finer samples experience larger vibration, i.e. they are impacted by larger dynamic forces. It should be noted that respective dependencies on the value of current density show the inverse correlation. The pulses of equal current density in samples of larger diameter experience vibration of larger amplitude. Corresponding dependencies are shown in Figure 8.

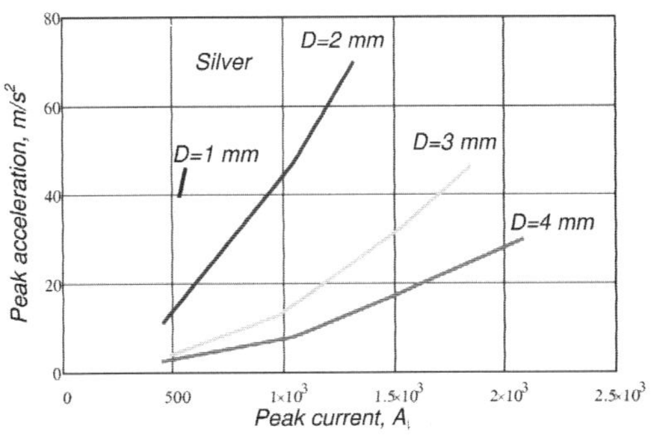

Figure 6. Dependency of peak acceleration on peak current for silver samples of different diameters.

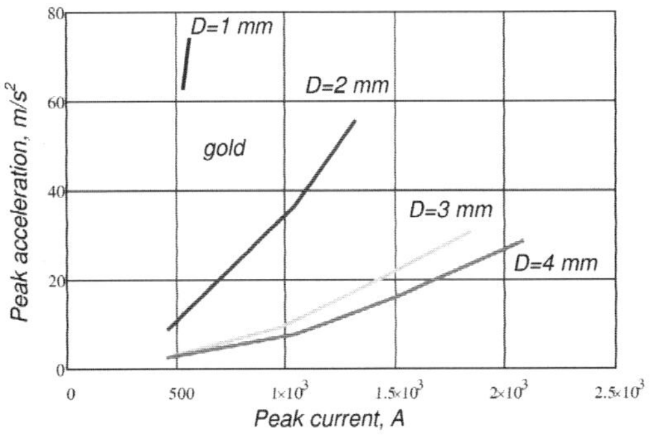

Figure 7. Dependency of peak acceleration on peak current for gold sampled of different diameters.

5. Analysis of the results

Results of experimental studies can be considered as confirming the model of the MHD interaction of pulse current in a conductor with its intrinsic magnetic field [11,14,15]. This applies to the time interval of the active exposure of the current pulse. After the action of the pulse the damped spatial vibrations correspond to the intrinsic vibrations of the fixed sample. With constant amplitude of the pulse current the peak value of vibration response is greater for the samples of smaller diameter, and for materials with a higher specific density. At the constant amplitude of current density the peak value of acceleration of vibration response is higher for the samples of smaller cross-section diameter.

During the active action of current pulse the peak values of amplitude of vibrational response acceleration are close to the moments of the maximum value of the derivative value of the magnetic field.

Vibrational response for the samples with higher specific density is larger, as shown in Figure 9.

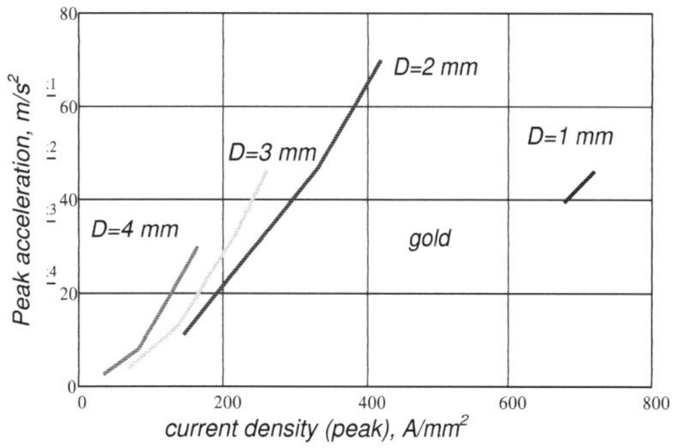

Figure 8. Dependency of peak acceleration value on the peak current density value in samples.

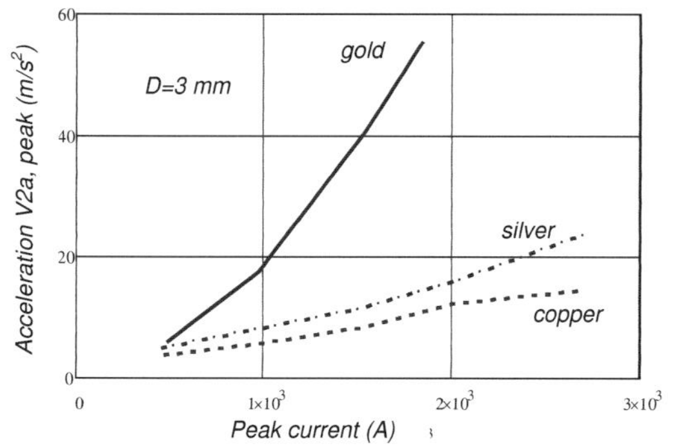

Figure 9. Dependency of peak radial acceleration on peak current transmitted through different samples of 3 mm in diameter.

6. Conclusion

The results indicate the complexity of the mechanism of excitation of mechanical vibrations in metal conductors by passing high-density current pulses. This must be considered in studies on the fatigue resistance of powerful electric equipment elements, in the development of metalwork methods with the use of the current exposure and in non-destructive test systems.

References

[1] Nasilovski J 1964 *Exploding wires* **3** 295
[2] Lukyanov A Molokov S 2001 Flexural vibrations induced in thin metal wires carrying high currents *Journal of Physics D - Applied Physics* **34 (10)** 1543-1552
[3] Troitskii O A 1984 The electroplastic effect in metals *Translated from Problemy Prochnosti* **2** 107–108
[4] Graneau P 1994 *Ampere-Neumann Electrodynamics of Metals* 311
[5] Darrigol O 2003 *Electrodynamics from Ampère to Einstein Electrodynamics from Ampère to Einstein* 559
[6] Baranov M 1992 Nonstationary electromagnetic and heating processes processes in cylindrical conductors under influence of lightning current *Electricity* **10** 9-15
[7] Frolova A Stolyarov V 2013 *Journal of Machinery Manufacture and Reliability* **42 (4)** 325
[8] Savenko V Troitskiy O Skvortsov O 2016 Influence of electro plasticity on the microstructure of steel under static load *Physical and engineering science and education: problems and research perspectives* 128

[9] Papageorgiou C Raptis I T E 2011 Collisions in metallic lattices under transient current pulses *Eur. Phys. J.: Appl. Phys* **54** 01
[10] Stepanov G 2006 Assessment of the effect of high-density electric current pulses on the plastic deformation rate of metals *Strength of Materials* **38 (2)** 183
[11] Surkaev A 2015 Magnetohydrodynamic perturbations arising in metallic conductors under the action of the discharge current *Technical Physics* **60** 7 p 981
[12] Martin H T Martin M L 2000 *LabVIEW for Telecom, Semiconductor, Automotive, Sound and Vibration and General Test and Measurement (National Instruments virtual instrumentation series)* 272
[13] Pravotorova E A Skvortsov O B 2015 Modelling of vibration tests of winding elements of power electric equipment *Journal of machinery manufacture and reliability* **44 (5)** 479
[14] Tskhondiya G A 2012 Simulating the effect of a high density electric current pulse on the stress field during plastic deformation *International Journal of Material Forming* **5 (2)** 157
[15] Skal, A S 2013 The full Lorentz force formula responsible for turbulence in solids and fluids and explained Faraday's paradox *International Journal of Scientific & Engineering Research* **4 (2)** 10

IOP Conf. Series: Materials Science and Engineering **218** (2017) 012021 doi:10.1088/1757-899X/218/1/012021

High-voltage electric pulse welding of magnetic cores made of rings of magnetic soft alloy 49K2FA

A V Yudin, E G Grigoryev, S S Bashlykov and V I Goltsev

The laboratory of advanced technologies for creation of new materials, National Research Nuclear University MEPhI, Kashirskoe shosse, 31, Moscow, Russia, 115409

E-mail: Artem.Yudin@icloud.com

Abstract. This research analyses the influence of the parameters of high-voltage electric pulse welding on microstructure of weld seams and mechanical parameters of samples. Special tooling was developed for the process of high-voltage electric pulse welding, allowing welding of ring samples. The method of determination of mechanical characteristics of toroidal samples was developed. An influence is established of parameters of high-voltage electric pulse welding, such as current density and applied pressure, on microstructure of a weld seam and density of a weld joint. The study of magnetic parameters of samples is conducted.

1. Introduction

Various types of welded joints and welding processes are currently used for production of solid devices. Conventional welding methods can not always meet the requirements of the welding joints. Consequently, the modernization is required, as well as development of new methods with wider potential [1]. One of the promising directions of welding is the high-voltage electric pulse welding.

A high-voltage electrical pulse passing through the welded parts causes a significant change in the state of the contacting surfaces. Intense alternating magnetic field and the applied pressure affect the blank details simultaneously with the ohmic heating in the zone of contact of details. This leads to a firm joint (good quality of welding) of details.

Depending on the initial state of the contacting surfaces, current pulse parameters and the applied pressure, both forming of a firm joint (at the optimal parameters values) and explosive destruction of the contact zone are possible [1]. Earlier studies [2] discovered that initial roughness of surfaces of welded parts has a substantial effect on the welding process. Experiments showed that the maximum strength of the T-type welding joints (core - disk) in the analyzed range is obtained at the value of roughness of surface (R_z) 20 μ to 40 μ. This achieves the maximum energy release and the maximum heating of the contact surfaces of the welded parts.

The high-voltage pulse welding method uses special high-voltage process equipment, which allows providing controlled pulse electric power impact on the joined parts. This equipment has been designed and created in the Interdepartmental laboratory of advanced technologies for creation of new materials of National Research Nuclear University MEPhI for implementing advanced electric pulse technologies of development of new materials and products made them, with unique structure and properties unattainable by other methods. A more detailed description of the equipment can be found in articles [3-5].

Modern mechanic engineering widely use various systems of electromagnetic energy conversion. The technical and operational characteristics of such systems are determined by the magnetic parameters of the permanent magnets and magnetic cores, made of soft magnetic material, in which capacity the 49K2FA precision alloy is used in Russian instrumentation engineering. The improvement of magnetic properties: an increase of saturation induction and the maximum magnetic permeability and a decrease of coercive force significantly affect the quality of the products, which use details made of this precision alloy.

The objective of this paper is to investigate the influence of technological parameters of the high-voltage electric pulse welding (the current pulse amplitude and the applied pressure) on the joint efficiency of the elements in the reference samples, compare the characteristics of magnetic circuits made of thin rings of soft magnetic alloy 49K2FA to the characteristics of the glued magnetic cores.

2. Experiment

A special tooling has been developed to carry out the high-voltage electric pulse welding process. The welding of toroidal samples of the alloy 49K2FA was conducted on the system shown in Figure 1a, with the tooling shown in Figure 1b. The high-voltage electric pulse welding system consists of a high-frequency pulse generator (1) and a executive unit (2). The technological tooling includes a brass rod (1), a central laminate liner (2) and a brass electrode punch (3).

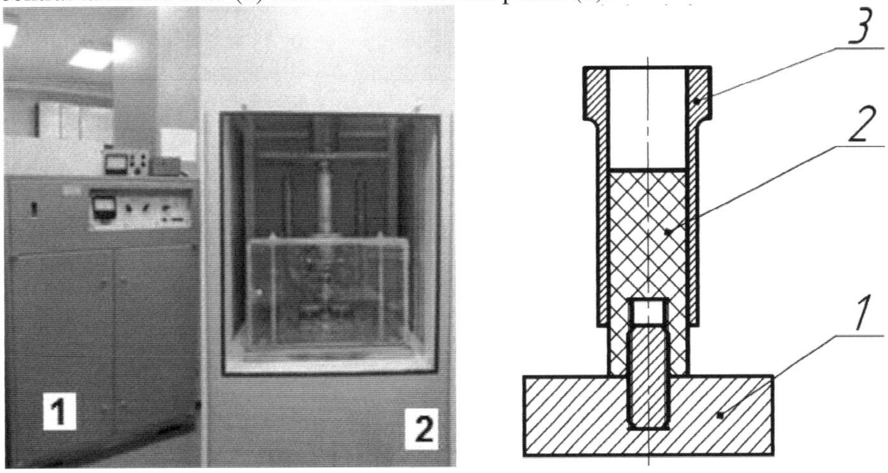

a b

Figure 1. Used equipment (a) and tooling (b).

Toroidal samples of 10 mm height were produced by welding at different technological modes of thin rings with thickness of 0,2 mm cut out of a sheet of 49K2FA precision alloy. In a high-voltage electric pulse welding process the main variables are the value of the ring (disks) compression force and the value of current density through the blank sample, as well as the number of pulses passed through the blank.

Changing the current density was carried out by changing the voltage on the capacitor bank, and changing the compression force was carried out by changing the pressure in the pneumatic system.

The testing is proposed of the ring samples by compression in a ring plane. Compression testing of a ring sample allows simplify the process of loading the sample and its destruction as fully as possible and determining not only the mechanical characteristics of the material, but also the quality and reliability of the joints of elements in cases when the produced sample is made of thin ring elements by the method of high-voltage welding. Essentially this realizes a lateral bending of a beam placeв on two supports (Fig. 2).

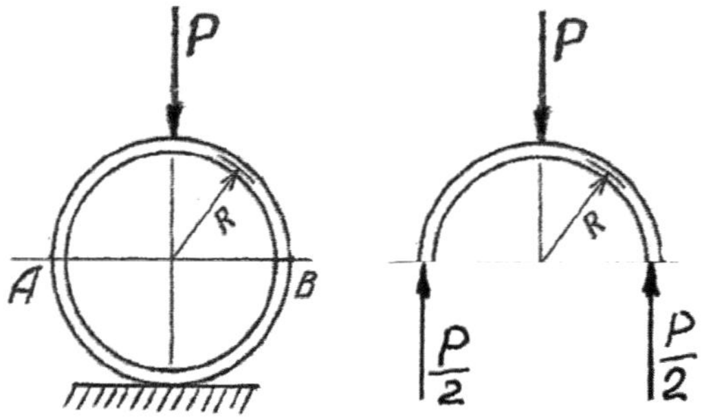

Figure 2. Schematic diagram of a compression test of the reference samples

Considered are three possible types of sample failure under a compressive loading by the scheme presented in Figure 2:

- brittle fracture on the plane of load application
- plastic deformation of the ring up to its flattening (without a fracture or with formation of cracks in the external side surfaces or internal along a plane coinciding with the plane of the loading);
- delamination of the produced ring on separate elements (exposed to weld joint) followed by loss of stability of these thin elements.

If the material of finished product is brittle, than the sample failure occurs along the plane of load application, where the maximum bending moment applies, bending strength is determined by the formula:

$$\sigma_B^u = \frac{3P_k D_{ad}}{2bh^2} \tag{1}$$

where P_k – recorded loading at the sample failure;
D_{cp} – average ring diameter;
b – ring width;
h – ring thickness.

Expected significant variation of the test results is typical for brittle fracture and may require a statistical processing.

If the material is plastic, a gradual flattening of the ring occurs with a possible·formation of cracks on the external or on the internal surfaces of the ring. In this case the deformation (the deflection of the sample) at the moment of the crack appearance is determined as well a characteristic of the strength of the material resistance to the initial appearance of plastic deformation – the yield bending strength, determined by the formula:

$$\sigma_T^u = \frac{3P_T D_{ad}}{2bh^2} \tag{2}$$

where P_T – the value of loading, corresponding to load corresponding to the maximum relative deformation of the sample 0,002 (determined by the diagram of loading-deflection of the sample);
D_{ad} – average ring diameter;
b – ring width;
h – ring thickness.

If during the loading of the ring sample delamination and subsequent loss of stability of the individual elements occurs, it does not mean that the strength of material joint in the weld is insufficient. It is important to establish at which stage of loading the delamination occurs. Ensuring the strength of the sample and the reliability of the element joint is achieved under optimal conditions of high-voltage welding.

3. Experimental results and discussion

A study of microstructure of the sample produced by a single pulse of 2 kV and the pressure of 88.5 MPa, revealed a rather large gaps between the rings. The weld seam is a spot weld of size about 50 μm (see Fig. 3 a). Conducting mechanical test allowed determining that the sample is degrading by delamination of the finished product into the separate elements with a consequent loss of stability of these thin elements. The delamination of rings occurs at the pressure equal 43.2 MPa and the displacement value equal 0.17 mm (Fig. 3 b).

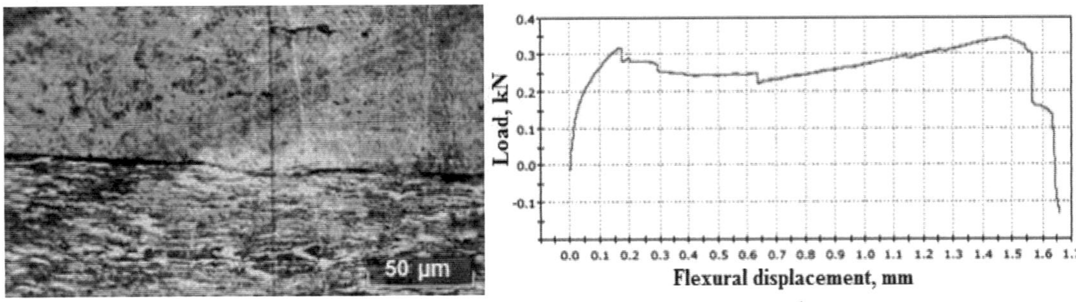

a b

Figure 3. Characteristics of the sample produced by a single pulse of 2.0 kV

The microstructure of the sample welded by a single pulse of 3.5 kV at the pressure equal 88.5 MPa is characterized by smaller gaps between the disks, at the same time the size of the weld seam remained nearly the same (about 50 μm). Mechanical tests showed that the pressure value at the sample failure amounts to 67.5 MPa, and deformation equals 1.1 mm. The character of the destruction also changed with no delamination of the rings and it proceeded by the scheme of plastic deformation of the ring.

Next samples were made by two high-voltage pulses of 3.0 kV at the pressure of 88.5 MPa. At these parameters of the high-voltage electric pulse welding the gaps between the disks are practically nonexistent and the size of the weld seam amounts to 70 μm to 100 μm (Fig.4, a), and the failure pressure increased to 185.5 MPa, with the maximum deformation equal 2.6 mm. The failure also proceeded by the scheme of plastic deformation of the ring, with no sharp declines of load value leading to the delamination of the rings from the sample (Fig.4, b).

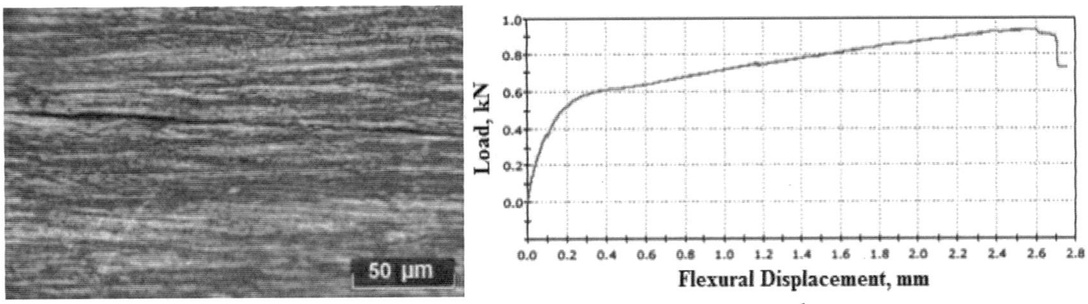

a b

Figure 4. Characteristics of the sample produced by two pulses of 3,0 kV

Next set of samples was produced by passing two high-voltage pulses of with an amplitude of 4.0 kV at the pressure equal 88.5 MPa.

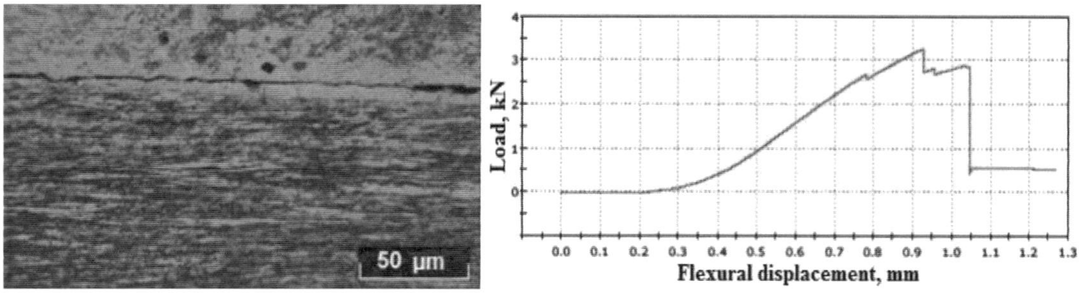

a b

Figure 5. Characteristics of the sample produced by two pulses of 4.0 kV

As is evident from the photograph of the microstructure there are no gaps between the discs and the size of the weld seam amounts to more than 100 μm (Fig. 5 a). Concurrently the ultimate strength amounts to 351 MPa, the deformation equals 0.77 mm, but the character of the failure has changed. In this set of samples the brittle failure is prevalent (Fig. 5 b).

Consequently, the overall analysis of sample loading diagrams and recorded stepwise images of the process leads to the conclusion that the delamination of ring samples starts outside of the elastic initial loading zone and is reflected in the diagram in the form of steps.

The study of magnetic properties of the samples showed that with this method of joining rings into the magnetic core there is no deterioration of the magnetic characteristics of the material. E.g. for the sample produced by two pulses of the amplitude 4.0 kV, the following magnetic characteristics were obtained: $B = 2.26$ T, $H_c = 50$ A/m, $\mu = 15000$ G/Oe.

4. Conclusions

The research proposes a method of strength testing of weld seam quality. Such strength testing allows simplifying the process of loading the ring sample and its failure as fully as possible and determining not only the mechanical characteristics of the material, but also the quality and reliability of the joints of elements. The comparison of the microstructure and strength properties of welded joints showed the following:

- the strength the weld joint depends on the distance between the plates in the sample, the shorter the distance the greater the strength;
- the use of two high-voltage pulses increases the weld zone (weld seams), which also increase the strength of joint;
- the location of weld seams between the plates depends on the roughness of the welded surfaces, which define the contact zone
- An optimal mode of obtaining quality weld seam is suggested: two high-voltage pulses with a voltage value of 4,0 kV.

References

[1] Grigoryev E G 2011 High voltage electric pulse welding of titanium with stainless steel *International Scientific Journal Machines. Technologies. Materials* **8** 8-11
[2] Gurevich S M 1990 *Reference book on welding of nonferrous metals* (Kiev, Naukova Dumka)
[3] Grigoryev E G, Bazanov V N 2008 Electric current pulse welding of ti*tanium VT1-0* with stainless steel 12X18H10T *Moscow State University of Education J.: Natural* Sciences **2** 84-88
[4] Grigoryev E G, Bazanov V N 2010 Electric current pulse welding of titanium with 18-10 STAINLESS STEEL *Advanced Materials Research* **83-86** 1251
[5] Grigoryev E G 2011 Welding of titanium with stainless steel by high-voltage electric-current pulse *Proceedings of the 2011 International Conference on Powder Metallurgy and Particulate Materials, "PowderMet 2011"* 642-646

Technological features manufacture samples of stainless steel 316L by Selective Laser Melting on the machine Melt Master3D-550.

A V Yudin, V V Beregovsky, E V Tret'yakov, J E Jhirnova and M A Burmistrov

Laboratory additive technologies, RPA "CNIITMASH", Moscow, 115088, Russian Federation

E-mail: Artem.Yudin@icloud.com

Abstract. In this work we consider obtaining high-density powder samples from austenitic stainless steel 316L, on the first national experimental layered build-up equipment by the method of selective laser melting (SLM) Melt Master3D - 550. The equipment description is given. Morphometric properties of the original powdered material is studied. Continuous unit vectors (tracks) and high-density samples with the relative density of 97% are obtained during the experiments. The laser radiation energy contribution is estimated to compare the conditions for obtaining samples with different technological parameters. In course of the study of the sample properties the dependence of density on the specific energy is revealed. Moreover, the influence of technological parameters on the microstructure and microhardness of the samples is established. The obtained data are compared with the results of other studies on similar foreign facilities.

1. Introduction

At present, high-technology production trends of such industries as machinery, electric energy (including nuclear), oil and gas, medical show great interest in additive technologies of making products with complex shapes from metal and ceramic powder materials. Additive manufacturing with the selective laser melting (SLM) is a layer-processing technology by local melting of metal powder materials with laser energy, which makes it possible to obtain high-quality products with complex shapes immediately, i.e. the subsequent final processing in most cases can be completely excluded.

Over the past decade of the development of additive technologies a large number of research papers which describe obtaining samples and products from various types of stainless steel with the selective laser melting have been published. As a rule, the authors give the following main technological parameters that affect the quality of the obtained products and samples [1, 2]:

- original metal powder: chemical composition, shape and size distribution of particles, thickness of the melted layer for each production cycle;
- laser radiation source: type, capacity, space-energy beam parameters (radiation intensity distribution, the size of the laser beam spot in the melting zone as a result of its divergence);
- characteristics of the production process: scan speed, type of a protective gas environment, choice of the type and parameters of the scan strategy on each melted layer.

In the papers [3-5] the samples of stainless steel 316L are studied. The experiments were carried out on the selective laser melting (SLM) Phenix Systems PM100. The source of the laser radiation in this equipment is the ytterbium fiber laser YLR-50 CW at the wavelength of 1075 nm and a maximum output power P = 50 W. The size of the laser spot on the surface of the powder layer was about 70 microns. SLM was carried out in a closed chamber filled with nitrogen to prevent oxidation. The operating temperature in the chamber was kept at 80 °C.

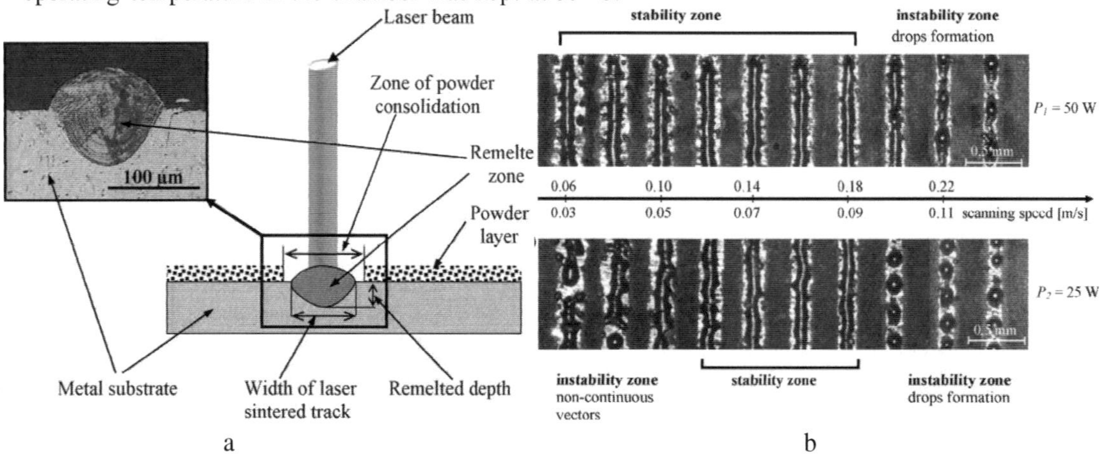

Figure 1. Tracks: (a) - cross-section of the track [4]; (b) - melted tracks from austenitic stainless steel [2].

The authors of the papers [3,4] describe the traces on the of stainless steel building platform at the laser power of 50 W and 25 W at various scan speeds in the range from 0.06 to 0.24 m / s and from 0.03 to 0.12 m/s respectively. The thickness of the powder layer was 50 microns. Figure 1 (a) shows a cross sectional view of the track indicating width, consolidation zone and the depth of melting zone. Figure 1 (b) shows photographs of the tracks made in different technological modes.

The authors of the paper [5] obtained bulk samples by combining the methods of "two zones" and "cross-hatch" with the following parameters: the laser power of 50 W, layer thickness of 40 microns, the scan speed of 0.12 m / s, the hatching step of 120 microns. The results of the mechanical tests of samples are shown in Table 1. The analysis of the mechanical properties of the samples produced by the SLM method from the powder of stainless steel (SS) 316L showed excellent mechanical properties matching cast materials.

Table 1. Mechanical properties of the SLM samples [5].

Properties of the sample	Way of the sample production	
	SLM	Casting
Tensile strength (MPa)	436±60	480-560
Yield strength (0.2%) (MPa)	366±50	170-290
Elongation (%)	9±2	40

The above-mentioned papers show the possibility of obtaining products from powders of austenitic stainless steels by SLM method on the basis of Phenix Systems PM100, Concept Laser GmbH M1. This article considers the production of high-density samples from stainless steel 316L on the first Russian equipment of layer build-up with the method of selective laser melting Melt Master3D550.

2. Experiments

The experiments were carried out on MeltMaster3D550, which external design is given in Figure 2. The main specifications of Melt Master3D550 are shown in Table 2, which, for comparison, also shows the parameters of similar imported facilities. In our experiments stainless steel 316L powder was used. Figure 3 shows photographs of the powder particles. The figure shows that the particles have a rounded shape. The estimates of the index of the powder particle geometry were obtained; the estimated average value was 1.5 with a standard deviation of 0.5. The particle size of the powder was in the range of 10 - 90 microns and the average particle size was 40 microns.

Table 2. Main characteristics of the facilities of different manufacturers producing layered products.

Equipment	SLM 500HL	EOS M400	X line 1000R	MeltMaster^{3D}550
Maximum sizes of the building platform (mm)	500×280×325	400×400×400	630×400×500	550×450×450
Laser, number, pcs. / Power, kW / type	2×0,4 (2×1,0) ytterbium fiber	1 / 1,0 / ytterbium fiber	1 / 1,0 / ytterbium fiber	1 / 1,0 / ytterbium fiber
Building rate (cm^3/h)	up to 70	-	10-100	15-100
Layer thickness (μm)	20-200	-	30-200	20-250
Spot diameter (μm)	80-150/700	90	-	50-700
Scan speed (m/s)	up to 15	7,0	-	up to 15

The following technological parameters of the SLM process for obtaining sample were used: the scan speed from 100 to 1000 mm / s and laser power from 154 to 491 watts. The samples were made as a rectangular parallelepiped with dimensions of 3x3x15 mm. The thickness of the layers during the sample building was 200 micrometers. The same hatch was made for all the samples with a simple one-pass strategy. The selective laser melting process of the samples occurred in a nitrogen atmosphere, residual oxygen at the concentration less than 0.07%.

Figure 2. External design of MeltMaster^{3D}-550.

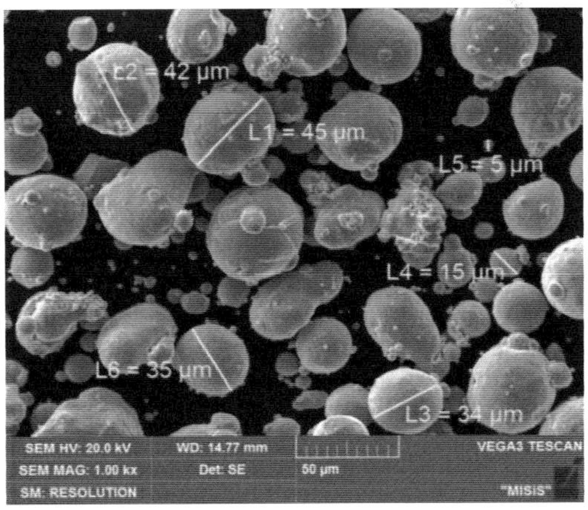

Figure 3. Morphological characteristics of the steel powder particles 316L.

3. Results and discussion

3.1 Measure of the track width

During the building of bulk samples one of the important parameters is a hatching step. To determine the right step for a given layer thickness the continuous tracks were obtained from a single layer of the same thickness as that of the samples (200 microns). To obtain the tracks the main parameters were the following: power from 379 up to 530 W and scan speed from 60 to 100 mm/s. The width of the tracks was measured by the microphotos. The hatching step was calculated on the basis of the following equation (1):

$$h = 0,7 \times b \qquad (1)$$

where h – hatching step, b- track width.
The calculated value of the hatching step is 180 microns.

3.2 Dependence of the sample density on SLM parameters

For comparing the conditions of sample building with different technological parameters, it is necessary to determine the contribution of specific laser energy. The sample density measurement results made under different technological SLM modes are shown on the diagram of the dependence of sample density on the specific energy (Fig. 4).

The evaluation of the specific laser energy for each tested sample was carried out according to the following formula (2) [6]:

$$E_{sp} = \frac{P}{V \cdot d} \tag{2}$$

where P - laser radiation power, V - scan speed, d - spot diameter.

The analysis of the dependences of sample density on specific energy showed that when the energy changes from 1 to 3 J/mm^2 the density increases dramatically, but at the same time the density value is less than 90%. When increasing the specific energy from 3 to 12 J/mm^2 there is an increase in density, but with a less dependence. The further increase in specific energy does not lead to the increase of the sample density actually. It can be also concluded that the optimum value of the specific laser energy is 8 - 15 J/mm^2. These values are consistent with the published data in the above-mentioned papers, and have a similar dependence of the sample density on the introduced specific energy [7]. When the samples are separated from the building platform the following result was reached: a group of samples made in the range of the input specific energy from 0.8 to 1.9 J/mm^2 was not welded to the platform.

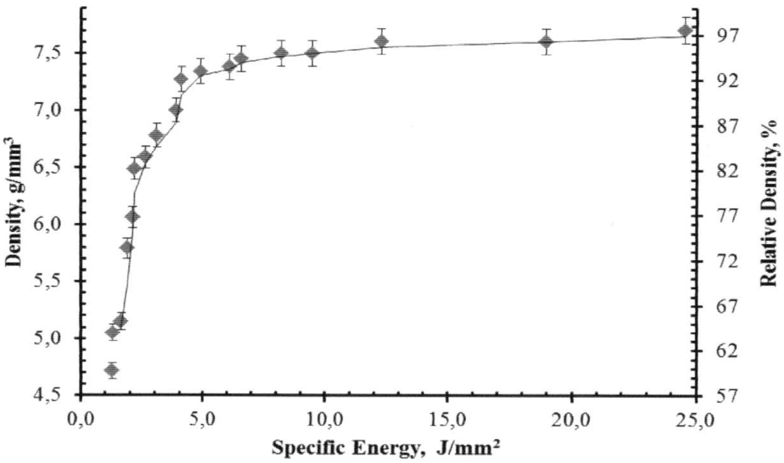

Figure 4. Dependence of the sample density on the input specific energy.

3.3 Microstructure Research

The results of microstructure studies of samples made with specific energy up to 3.0 J/mm^2 showed the presence of open macropores (shells) of more than 150 microns (Figure 5 (a)). For samples made with the input specific energy from 3.0 to 8.0 J/mm^2 pores of a 100 micron were typical. Furthermore, the the samples produced by the SLM method have a layered structure. The samples made with the input specific energy from 8 to 12 J/mm^2 were characterized by the presence of micropores.

The samples obtained at the input specific energy from 12 to 25 J/mm^2 have micropores and tracks of repeated melting. The pore distribution in this case is random. Figure 5 (b) shows the typical microstructure of micropores and repeated remelting of the example of the sample obtained at the input specific energy of 19 J/mm^2. All the samples found multidirectional dendritic segregation and fine-grained area. This structure is due to the rapid cooling of the molten pool and directed crystallization during selective laser melting.

IOP Conf. Series: Materials Science and Engineering **218** (2017) 012022 doi:10.1088/1757-899X/218/1/012022

Moreover, the element composition analysis of the sample with a density of 95% from the theoretical density obtained at the input specific energy of 12 J/mm². The results of the studies showed that there was no oxygen and oxides in the sample indicating little (less than 0.07%) oxygen in the working atmosphere.

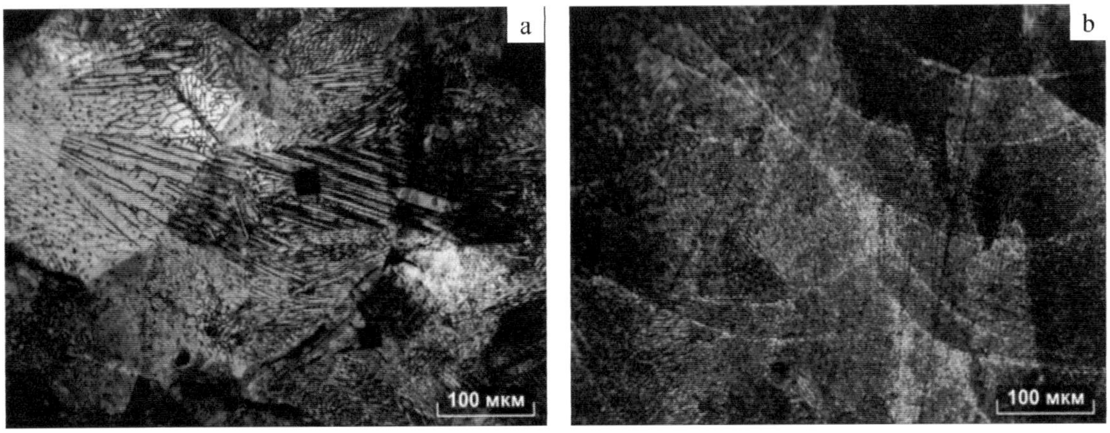

Figure 5. Typical microstructure of samples with macropores.

3.4 Microhardness

The measurement results are presented in the diagram of sample microhardness dependence on the specific energy of the SLM process (Figure 6). As seen from Figure 7 the microhardness of the samples obtained at the input specific energy of 8.2 J/mm² and 24.6 J/mm² is almost identical (the difference is less than 1%). This confirms that the optimal range of the value of the specific energy is 8 - 15 J/mm² for the layered synthesis process by SLM. Since the microhardness and density dependence on the technological SLM parameters is similar it can be concluded that the main reason of the sample microhardness increase with the increase of the specific energy of the powder melting is the decrease of the pore volume fraction.

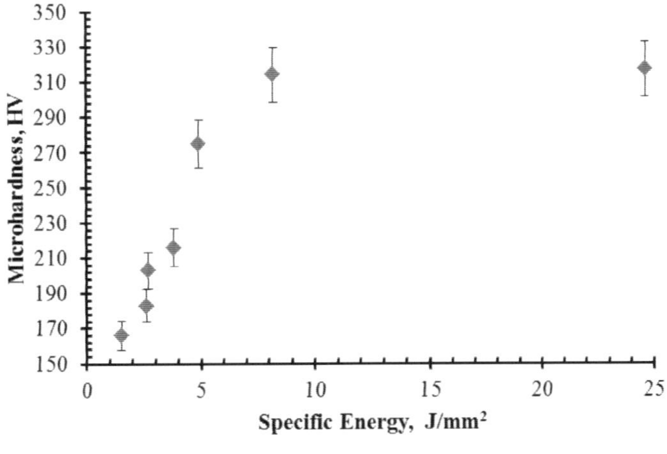

Figure 6. Dependence of microhardness on the parameters of SLM process.

4. Conclusions

The production of the first set of samples has shown the fundamental possibility for obtaining high-density samples from austenitic stainless steel 316L on layered selective laser melting MeltMaster3D-550 equipment.

In the course of continuous tracks study the right hatching step was calculated. At the thickness of the powder layer of 200 microns the right step between the adjacent passes of the laser must be 180 microns.

The study Identified dependencies of the sample density on the laser radiation power and scan speed. A more narrow range of input specific energy from 8 to 15 J/mm^2 is described.

The interrelation of the input specific energy and sample density has been set. The studies have shown that the dependence of the sample density on the input specific energy is consistent with the data presented in the paper [7].

The microstructural studies have revealed the dependance of selective laser melting parameters on the density, microstructure and microhardness of the samples. The samples from the steel 316L with the density more than 95% have the microhardness of about 315 HV.

Acknowledgments
This work was financially supported by the Ministry of Education and Science of the Russian Federation (The unique identifier PNIER RFMEFI58214H0004).

References

[1] Volosov MA, AA Okounkov Ways of optimization of selective laser melting process by selecting the laser beam machining strategies // Bulletin of Samara Scientific Center of the Russian Academy of Sciences. 2012. T. 14. №4 (2). Pp 587-591.
[2] Yadroitsev I., Bertrand P., Smurov I. Parametric analysis of the selective laser melting process // Applied surface science. - 2007. - T. 253. - number. 19. - P. 8064-8069.
[3] Kruth J. P. et al. Binding mechanisms in selective laser sintering and selective laser melting // Rapid prototyping journal. - 2005. - V. 11. - number. 1. - P. 26-36.
[4] Yadroitsev I. et al. Single track formation in selective laser melting of metal powders // Journal of Materials Processing Technology. - 2010. - T. 210. - number. 12. - P. 1624-1631.
[5] Smurov IJ, IA Movchan, IA Yadroytsev, Okounkov AA Cherkasova NY, Antonenkova GV Additive production with a laser. Conducting experimental studies // Vestnik MSTU "STANKIN". 2012. № 1. S. 36-38.
[6] Olakanmi E. O. Selective laser sintering / melting (SLS / SLM) of pure Al, Al-Mg, and Al-Si powders: Effect of processing conditions and powder properties // Journal of Materials Processing Technology. - 2013. - T. 213. - number. 8. - P. 1387-1405.
[7] Spierings A. B., Levy G. Comparison of density of stainless steel 316L parts produced with selective laser melting using different powder grades // Proceedings of the Annual International Solid Freeform Fabrication Symposium. - Austin, TX, 2009. - S. 342-353.